IMAGES
of America

LAWRENCEBURG

I0806368

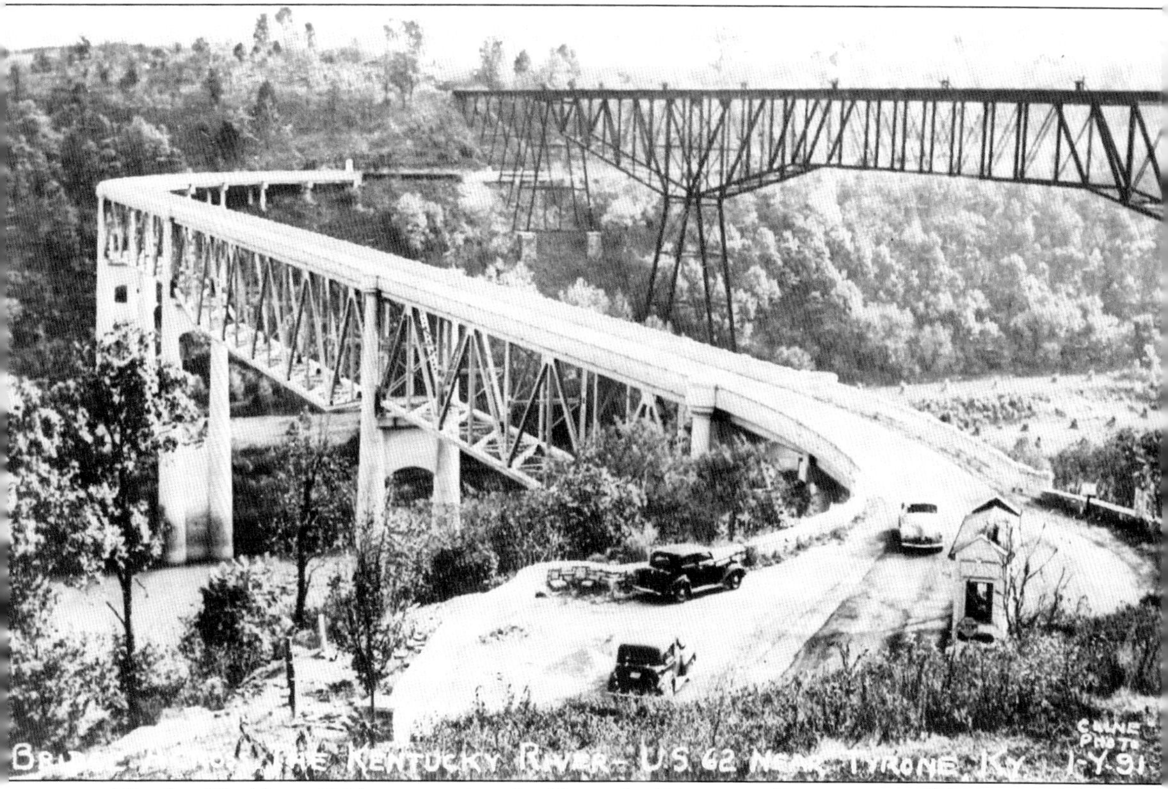

The Joe Blackburn Bridge, spanning the Kentucky River near Tyrone, opened for automobile traffic on June 11, 1932. Connecting Anderson and Woodford Counties, it is one of few bridges worldwide with two curves in it. Built at a cost of $480,000, the span was originally a toll bridge. A tool booth is visible in the lower right of the photograph. (Courtesy of McAnly collection.)

On the Cover: Lawrenceburg Supply Company, first begun by W.A. Bottom, had its main office on Court Street. Seen here from left to right are Cavel Harris, Earl Guthrie, Amos McKee, W.T. McCoun, Gertrude Thurman, and Charles McGinnis, the owner at the time. McGinnis was chairman of the Anderson County Industrial Corporation, which was instrumental in bringing industry to Lawrenceburg. (Courtesy of Kitty McCoun.)

IMAGES
of America

LAWRENCEBURG

William S. Bryant and Barbara S. Bryant

ARCADIA
PUBLISHING

ISBN 978-0-7385-9088-2

Published by Arcadia Publishing
Charleston, South Carolina

Printed in the United States of America

Library of Congress Control Number: 2012944089

For all general information, please contact Arcadia Publishing:
Telephone 843-853-2070
Fax 843-853-0044
E-mail sales@arcadiapublishing.com
For customer service and orders:
Toll-Free 1-888-313-2665

Visit us on the Internet at www.arcadiapublishing.com

We dedicate this book to our parents, Willard (deceased) and Elizabeth Bryant, Ezra Hubert and Catherine Sparrow (deceased), and our Aunt Vena (deceased), who instilled the values of family, friendship, and home in our lives.

Contents

ACKNOWLEDGMENTS

We would like to thank all of the people—family, friends, and old and new acquaintances—who were kind enough to share and trust us with their precious photographs, memories, and stories, truthful and otherwise, of Lawrenceburg's past.

We especially thank Bob McAnly, Brent Hawkins, and members of the Ripy and Geoghegan families for their extreme helpfulness in preparing this book. We hope that the book depicts the heart of Lawrenceburg, especially the people, places, and events that gave importance to our lives. We also hope it challenges others to gather more information and improve on the story we have told with words and images. We realize that we have barely scratched the surface.

Introduction

Lawrenceburg, the present county seat of Anderson County, Kentucky, was founded in 1820, seven years prior to the establishment of the county. Jacob Kaufman first settled the land more than 30 years before, and the settlement was named Kaufman's Station. Over time, it became known as Lawrence and then eventually Lawrenceburg.

The county was largely agricultural, and many residents grew tobacco and farmed, raised horses and cattle, and were dependent on bourbon distilleries to maintain the local lifestyle. The physical environment around Lawrenceburg, especially the topography, geologic and water resources, soils, and vegetation, determined the ways people could use the land during settlement times and throughout the 1900s.

A number of distilleries, many of which are award winning, dotted the local landscape and have maintained a strong presence over the years despite a decline during Prohibition. A number of distillers kept fine homes in town and maintained recreational cabins in the Kentucky River vicinity. Businessmen and other civic-minded citizens also contributed to the progress and stability of the community. A number of them served as elected city and county officials. Banks and bankers were especially influential in the town.

Although bourbon was a leading product in Lawrenceburg, many locals rejected it on religious grounds. Many denominations, including Presbyterians, Baptists, Christians, Methodists, and Catholics, present in the beginning of Lawrenceburg are still a big part of the fabric of the community today.

Lawrenceburg native son Champ Clark rose to national prominence as the speaker of the US House of Representatives and was a presidential contender in 1912. Clark began his career as a teacher in the county before switching to politics.

Originally a private academy before becoming the county high school, Kavanaugh School excelled in academics as a preparatory school for the US military academies. Rhoda Kavanaugh supervised Kavanaugh students for more than 40 years. Basketball teams and individual players earned honors for their heroics at Kavanaugh and Lawrenceburg High School. Forest "Aggie" Sale, a Kavanaugh graduate, became coach Adolph Rupp's first All-American basketball player at the University of Kentucky. Other athletes, especially the four-legged kind, were world-renowned in the Saddlebred horse industry.

Lawrenceburg's businesses and industries grew because of a well-developed transportation system—with early railroads and later highways—and its central location in the state, being within 100 miles of Lexington, Louisville, and Cincinnati. The fact that no local banks were forced to close during the Great Depression is a testament to the stability of the local economy.

The thread factory strike in the early 1950s was a setback to Lawrenceburg's growth, but the community kept looking forward. Today, Lawrenceburg functions more as a bedroom community to the surrounding urban areas than as a central hub.

One

SETTING

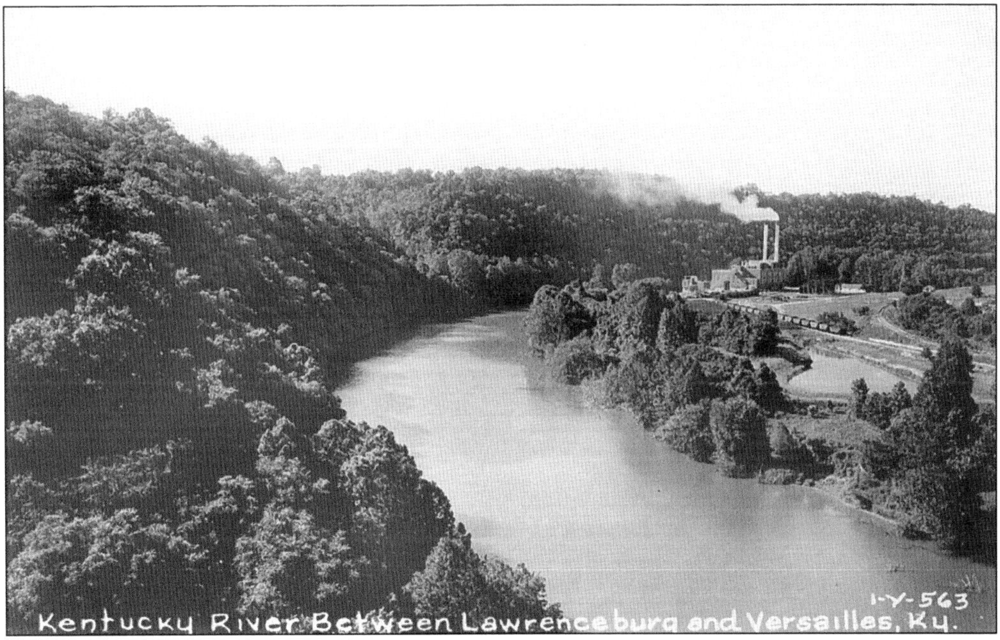

In 1784, John Filson listed the Kentucky River as one of the state's greatest peculiarities. The deeply entrenched river was formed by a rejuvenated stream that cut through Kentucky's dense limestone. Upland channels and meanders are still found within a few miles of the present river, which is 300 to 400 feet below its ancient course. (Courtesy of McAnly collection.)

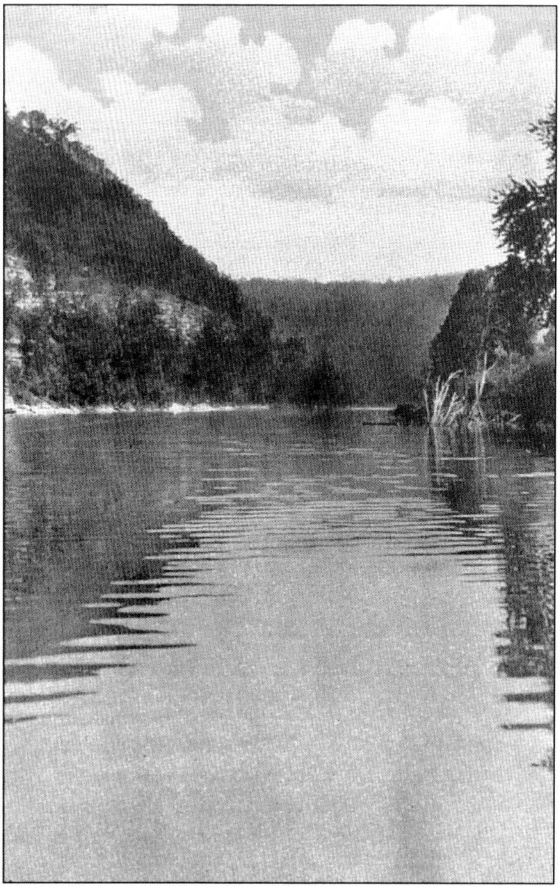

The Kentucky River begins at Pine Mountain and enters the Ohio River at Carrollton. The river rocks exposed on the cliffs are from the Ordovician Period and are approximately 460 million years old, the oldest in Kentucky. (Courtesy of McAnly collection.)

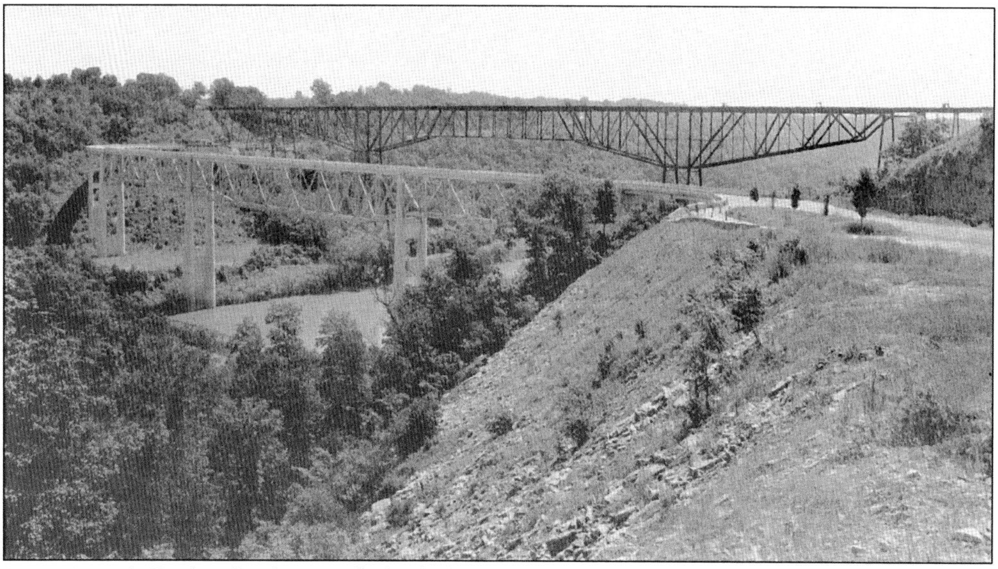

Young's High Bridge (background), built for the railroads, and the Joe Blackburn Bridge, for automobiles, both span the Kentucky River near Tyrone. Before these bridges, the Shryock Ferry was the only way to cross the river. (Courtesy of Anderson County History Society.)

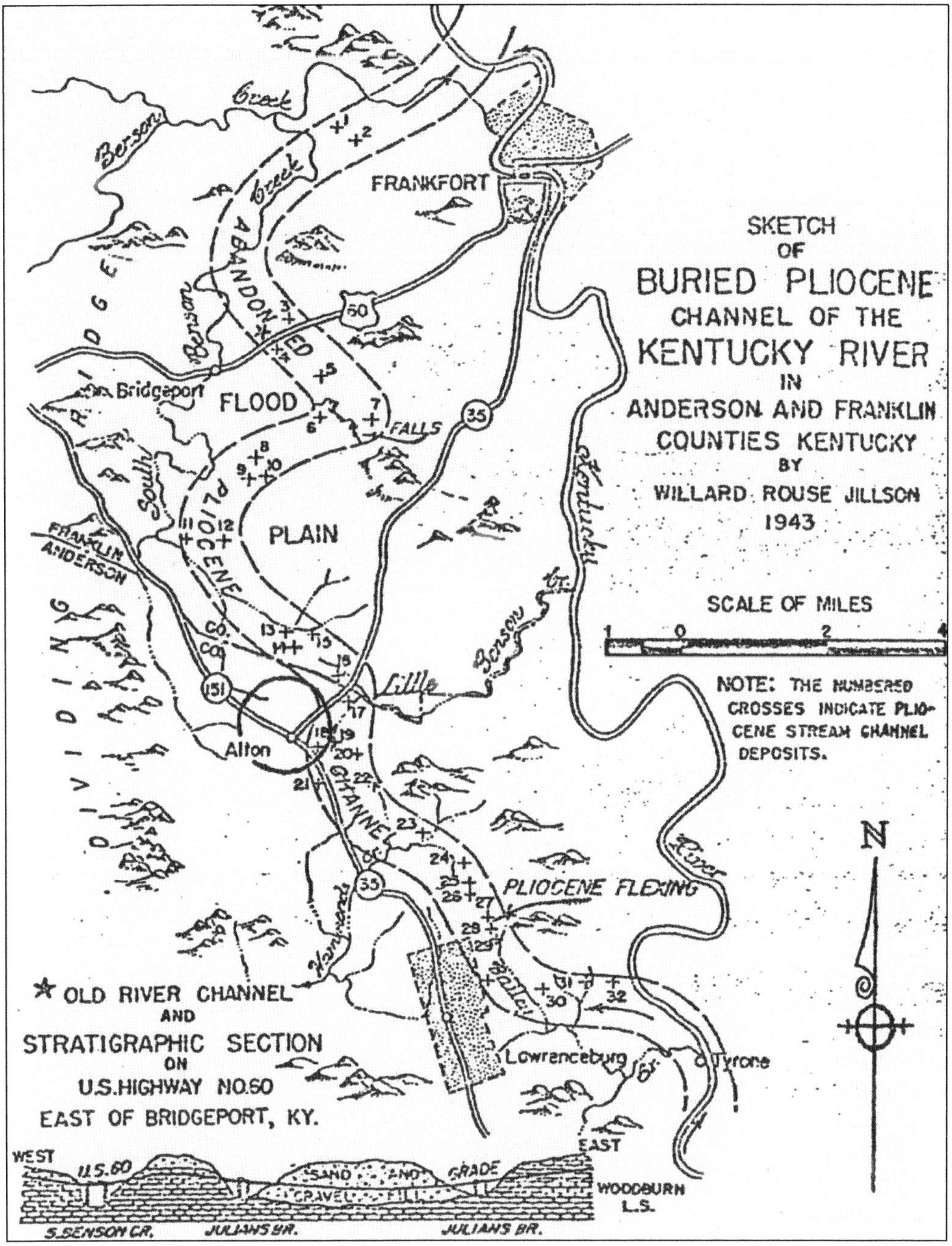

The Benson Abandoned Channel of the Kentucky River, seen here, shows that the river once ran on the upland, where it was a broad, shallow stream. Through the centuries, it was rejuvenated, changed course, and started cutting down to form the present channel. These channels were mapped by professor W.R. Jillson. (Courtesy of Bryant collection.)

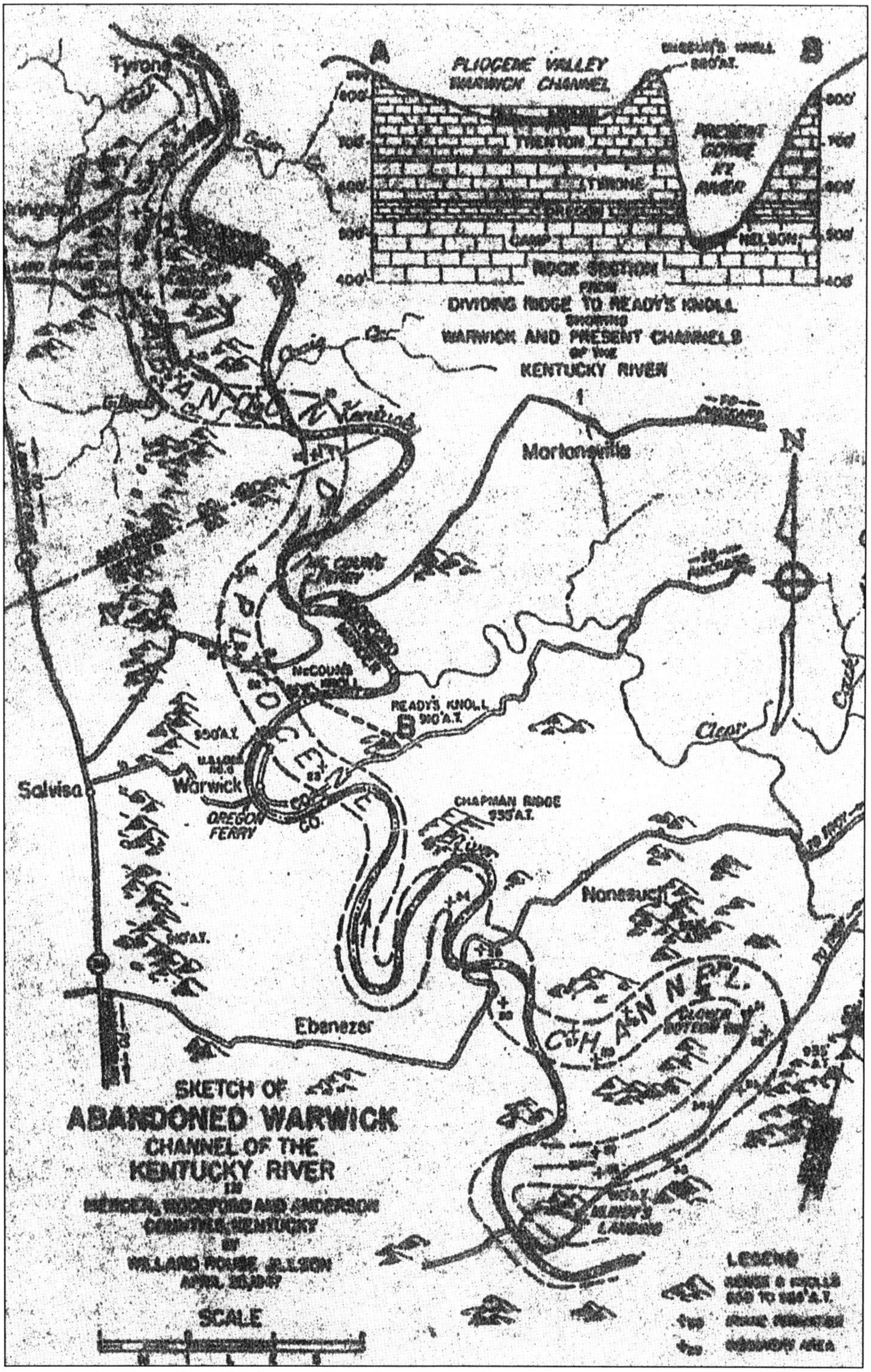

The Warwick Abandoned Channel of the Kentucky River is a continuation of the Benson Channel to the south. These upland channels feature deposits of Irvine gravel as well as geodes. (Courtesy of Bryant collection.)

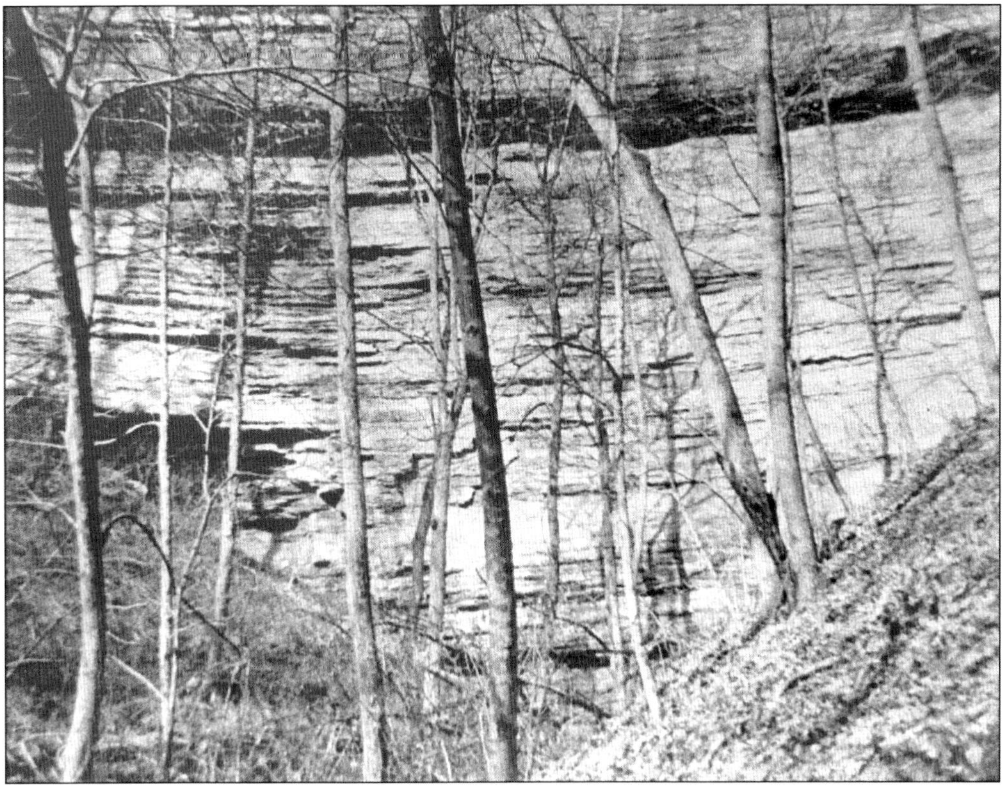

Panther Rock is a 110-foot-high fossil waterfall in the Wildcat Meander of the Warwick Abandoned Channel of the river. Water once flowed over the falls but later turned into a pond behind the falls, where solution formed the cave at the base of the falls (left). The solution and down cutting are still occurring today, exemplified by the spring below the cave entrance. (Courtesy of Bryant collection.)

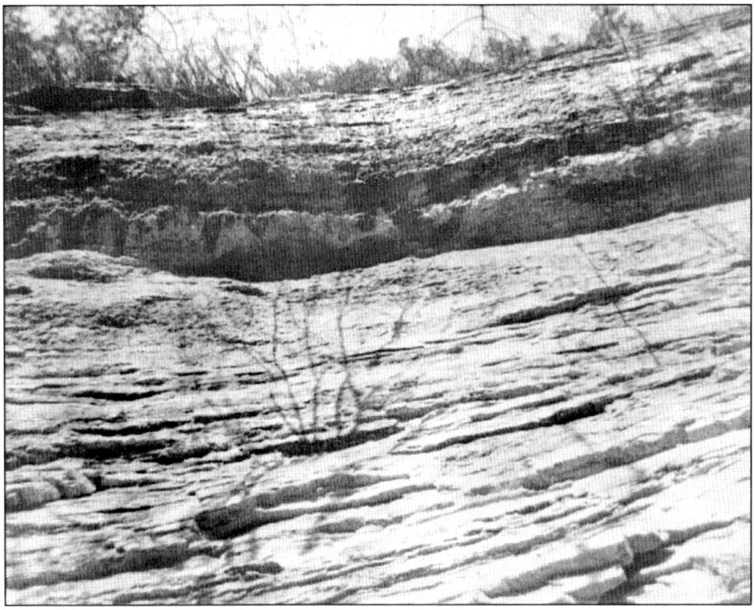

This lip rock of the Panther Rock waterfall shows why it is a fossil waterfall: water no longer flows over it. (Courtesy of Bryant collection.)

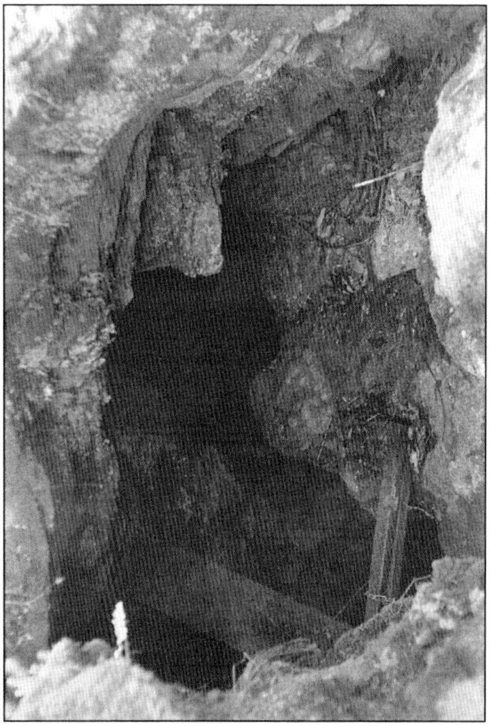

Caves, pit caves, and sinkholes dot the land where the Kentucky River formerly flowed. The limestone bedrock is soluble to water, and the many sinks and caves are evidence of its erosive power. (Courtesy of Bryant collection.)

McCalls Spring is one of many springs near Lawrenceburg. In 1773, the McAfees camped there as the first known white expedition into Anderson County. The McAfees were a surveying party headed by three brothers: Robert, George, and James. James McCoun Jr. and Samuel Adams were also in this party of adventurers. The bourbon industry's use of limestone springs in the production of whiskey has made the Bluegrass Region famous. (Courtesy of Anderson County History Society.)

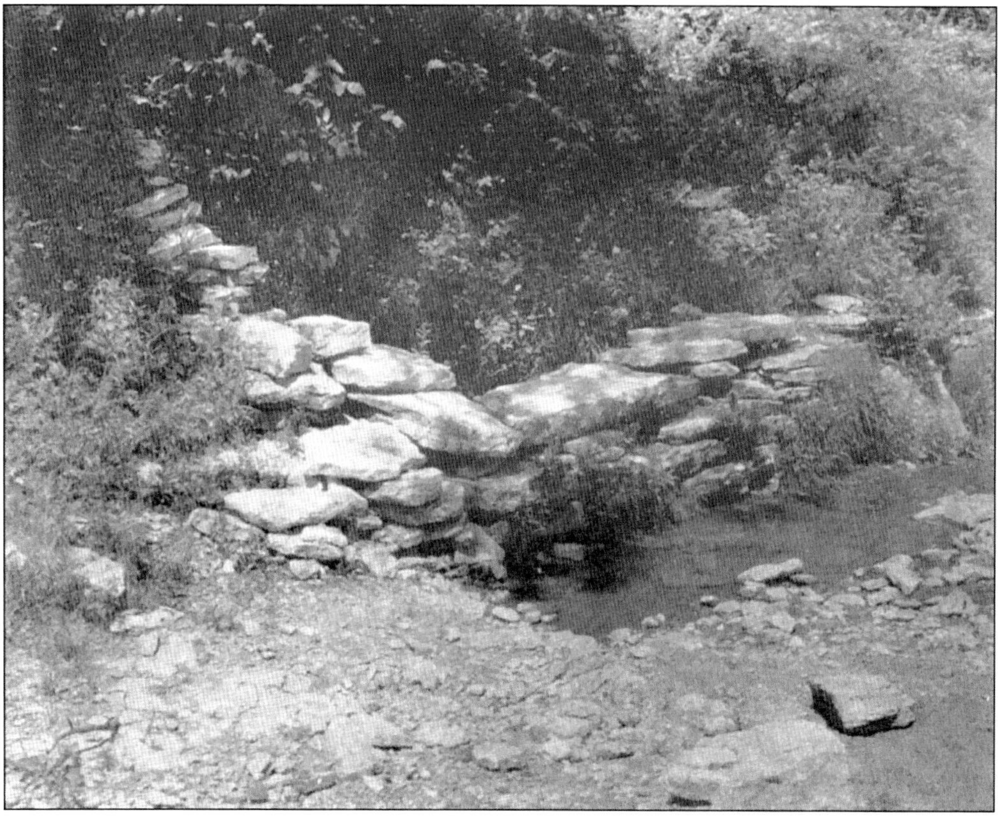

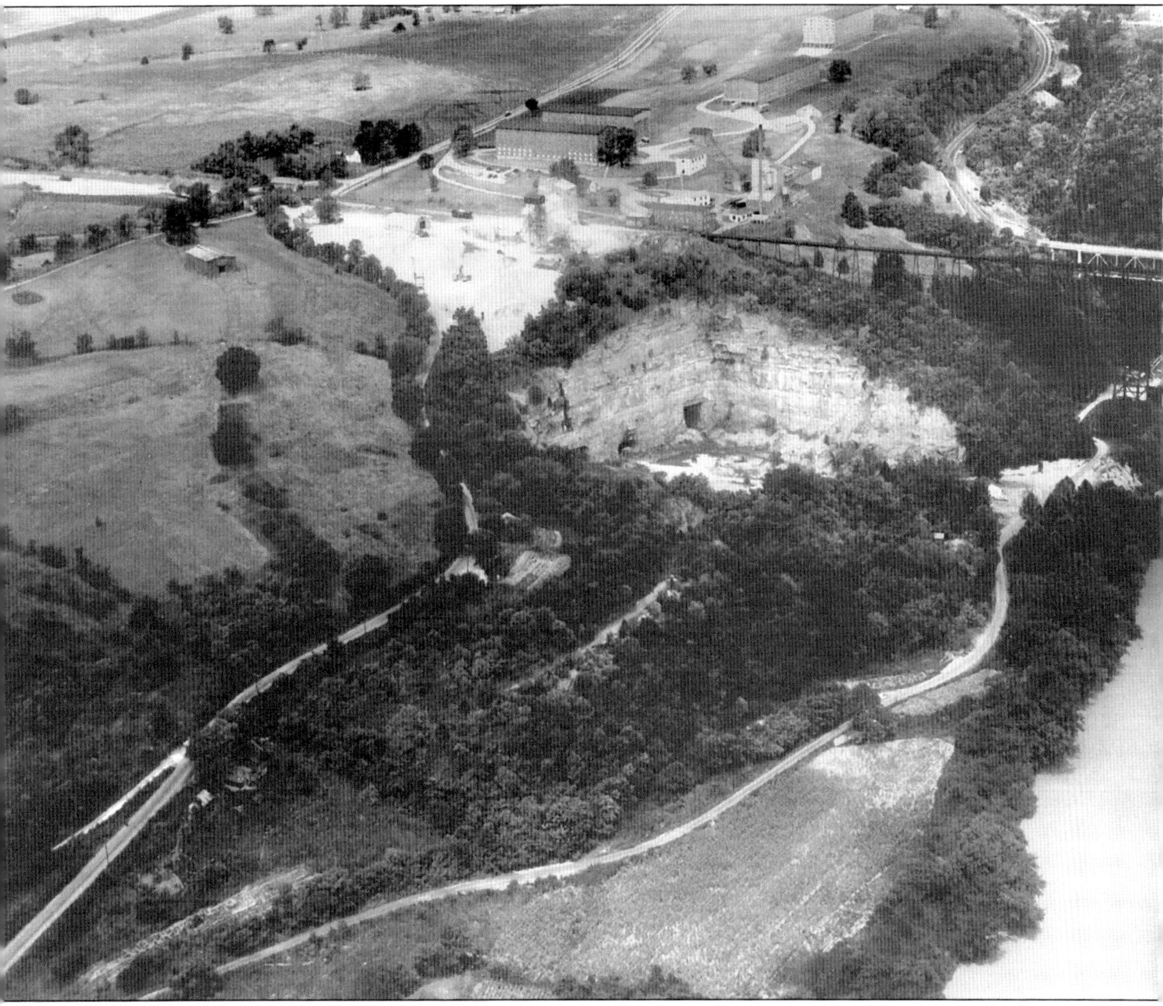

Limestone has been quarried for various uses. The Kentucky Stone Company's operation in Tyrone is seen here; note the tunnels near the base of exposed limestone. Crushed limestone is above the quarry. The Wild Turkey Distillery is to the right of the crushed stone, and the Blackburn Bridge crosses the Kentucky River to the far right. (Courtesy of Allen collection.)

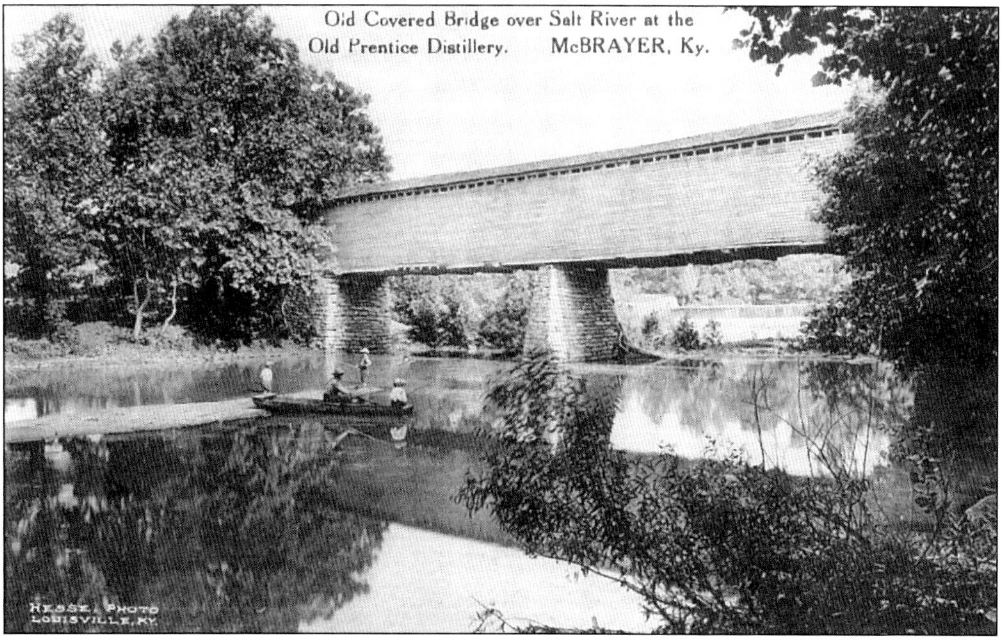

In the 1950s, the Kentucky River replaced the Salt River as Lawrenceburg's water source. The Salt River is a narrow and shallow stream that often slows to a trickle in dry months. This covered bridge, one of the many former covered bridges in Anderson County, at the Old Prentice Distillery (now known as Four Roses) spans the Salt River, which pirated Kentucky River near McBrayer in the ancient past. (Courtesy of McAnly collection.)

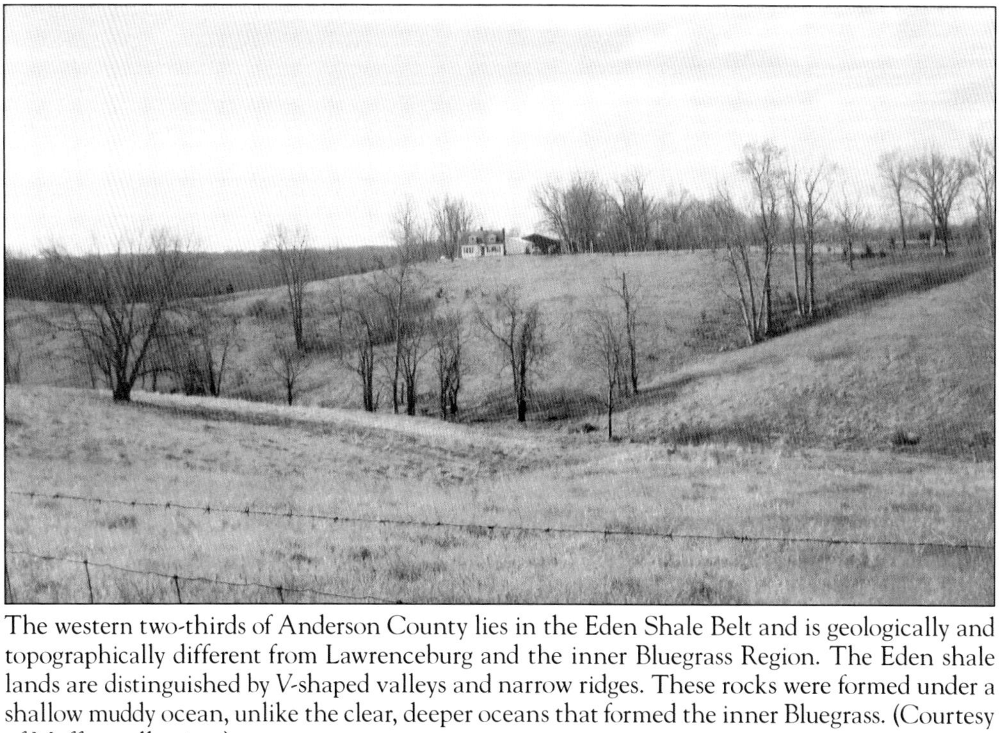

The western two-thirds of Anderson County lies in the Eden Shale Belt and is geologically and topographically different from Lawrenceburg and the inner Bluegrass Region. The Eden shale lands are distinguished by V-shaped valleys and narrow ridges. These rocks were formed under a shallow muddy ocean, unlike the clear, deeper oceans that formed the inner Bluegrass. (Courtesy of Moffett collection.)

Two

AGRICULTURE

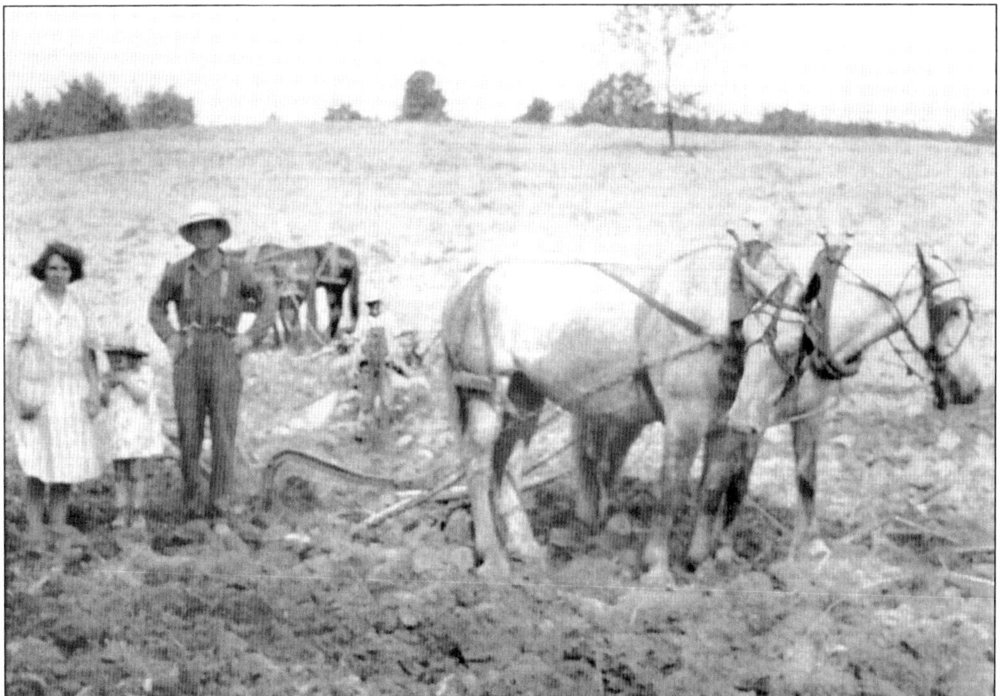

Guy, Nellie, and little Linda Moffett prepare a field for planting by plowing the lower slopes. Charles Best is in the background. (Courtesy of Moffett collection.)

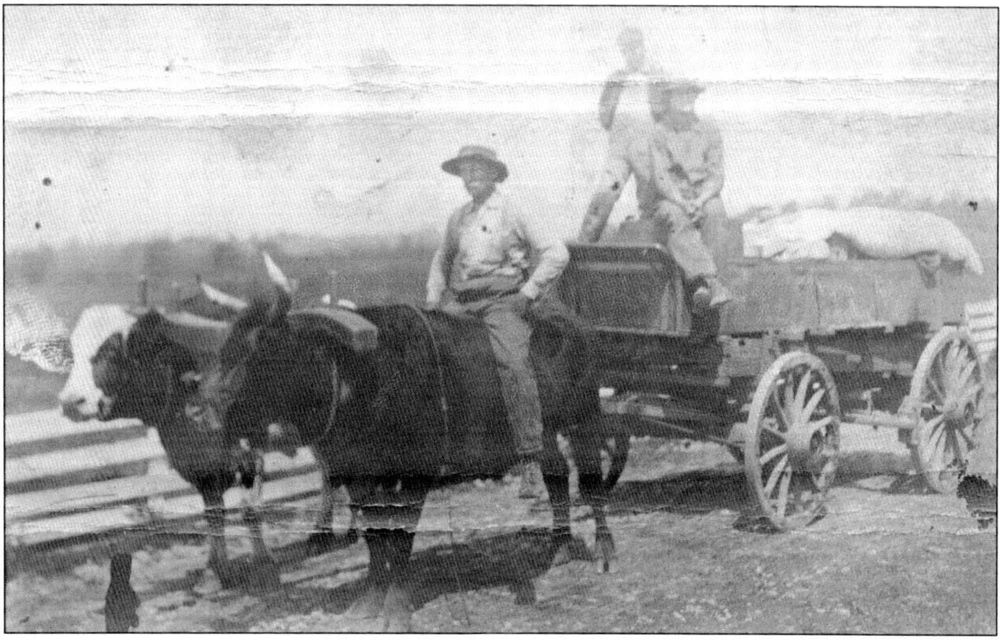

A team of oxen was used for heavy work on the farm. This team hauls sacks of crushed corn that was ground at the local mill. (Courtesy of McFarland collection.)

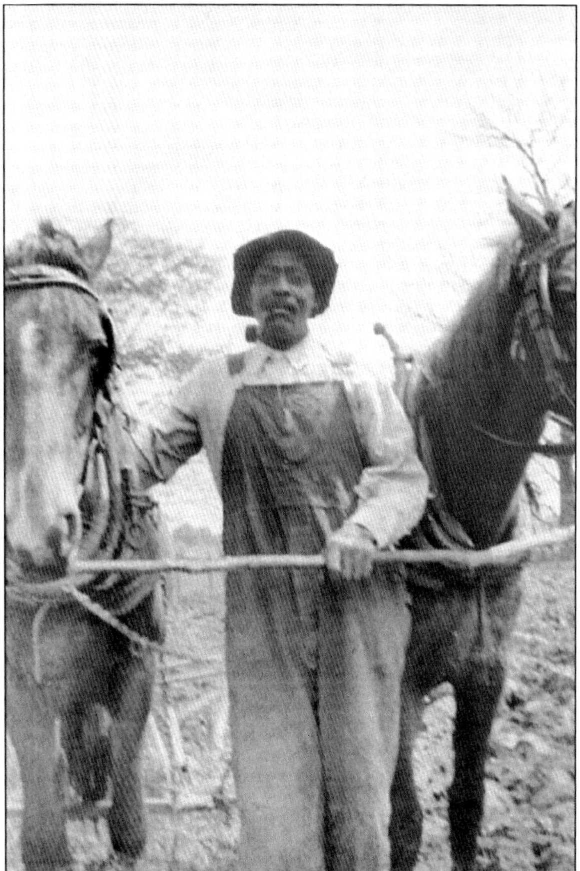

Jess Thurman leads a team of horses as he prepares to harrow a field on the Roscoe Bryant farm. (Courtesy of Bryant collection.)

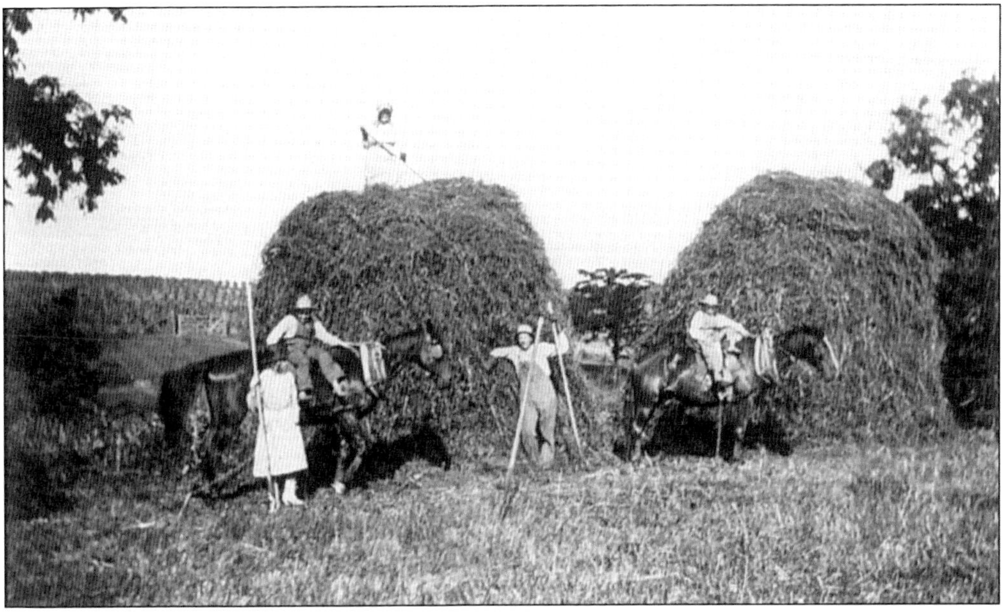

Men and women worked together on the Hanks farm. After hand-raking the hay, they placed it in stacks to be used later. (Courtesy of McFarland collection.)

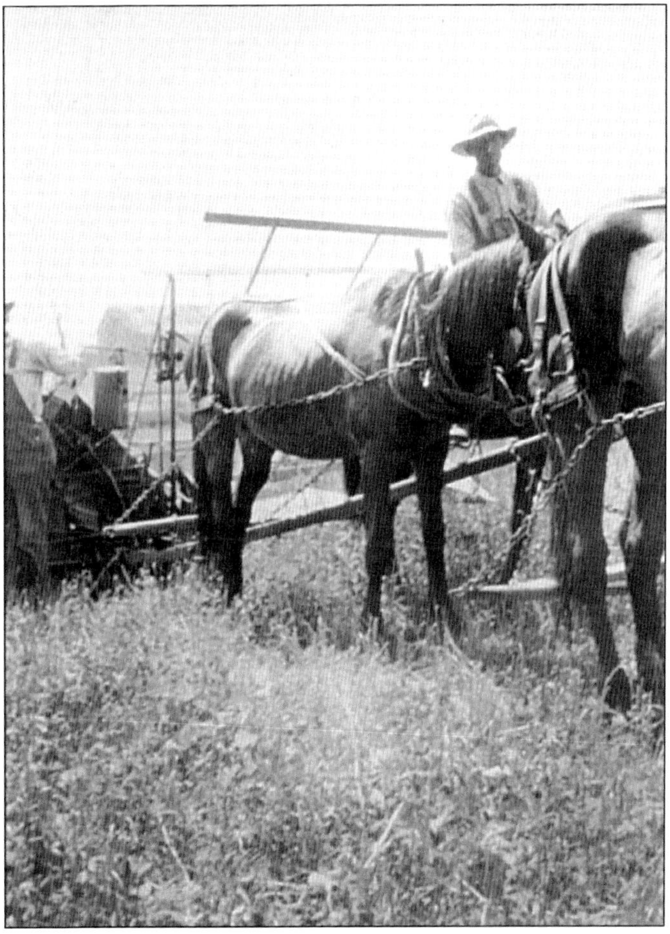

Bill Bryant rode a team of horses to harvest the grain. In the 1930s, the lack of rain was evident in stunted crops. (Courtesy of Bryant collection.)

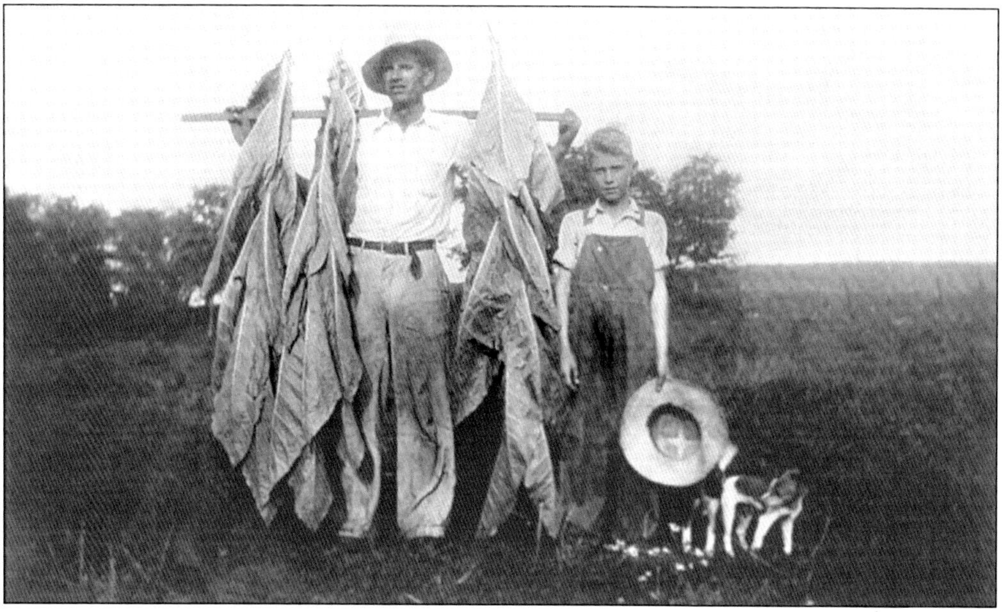

Porter Smith and his son W.J. pose with recently cut tobacco. Tobacco was the main cash crop, and a good harvest was essential since tobacco was sold by the pound. (Courtesy of McFarland collection.)

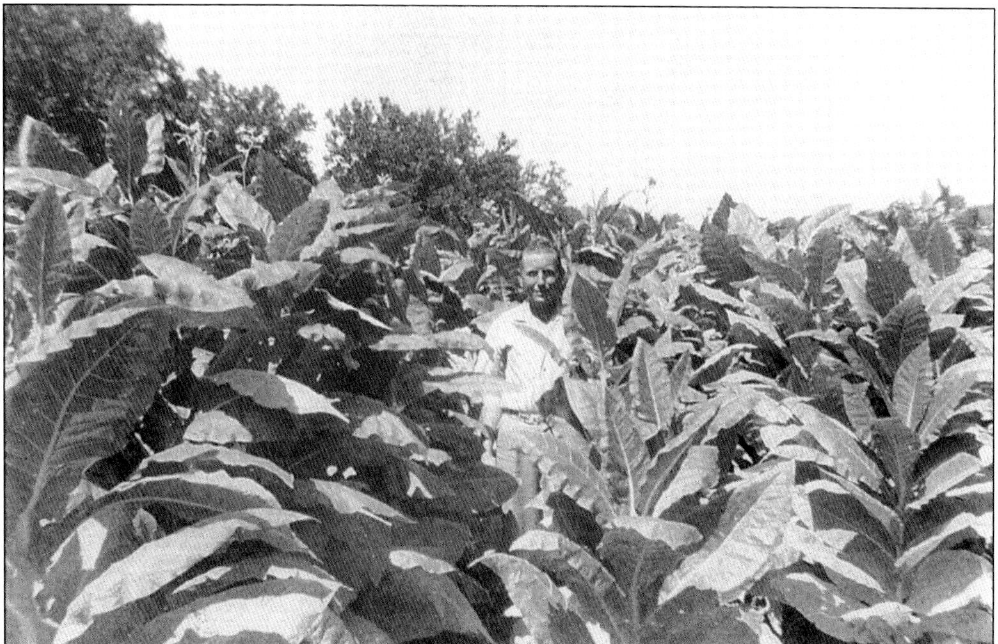

Hubert Sparrow is almost finished topping the tobacco in his crop near Stringtown. (Courtesy of Bryant collection.)

Roscoe Bryant cuts tobacco on his farm on Jenny Lillard Road around 1938. After being topped, the leaves expanded, giving the tobacco weight. Since it was sold by the pound, large-leaf tobacco was preferred. (Courtesy of Bryant collection.)

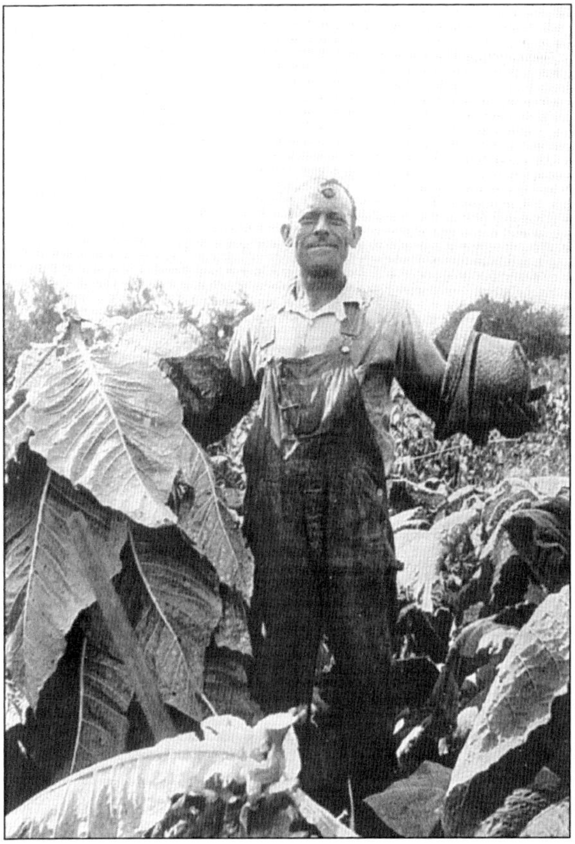

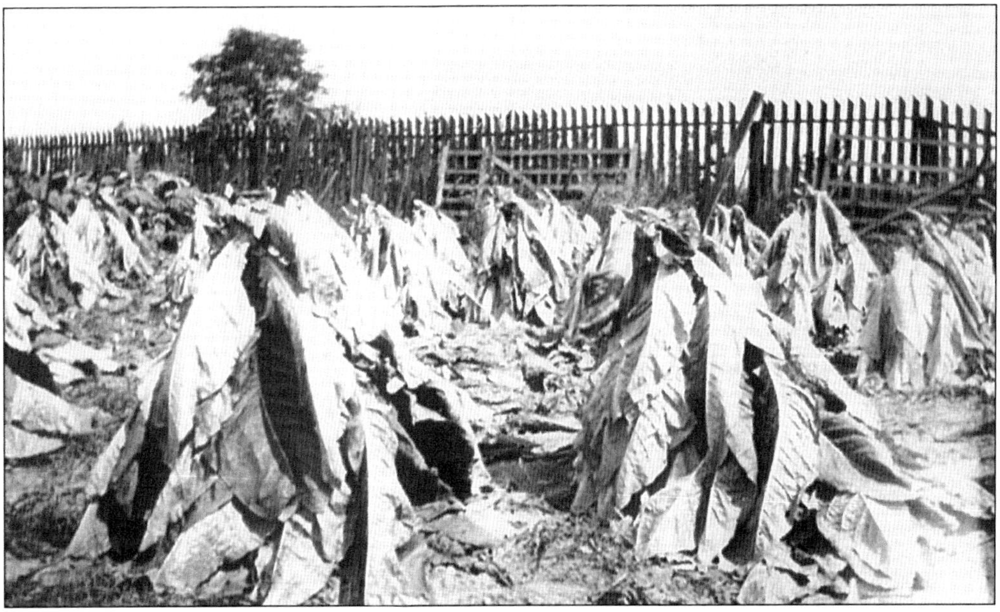

A field of cut tobacco is seen here. It was necessary for it to wilt and lose moisture so that the leaves did not break off. After the tobacco was cut, it was housed in a tobacco barn, where it was dried and cured before being handled again. (Courtesy of Bryant collection.)

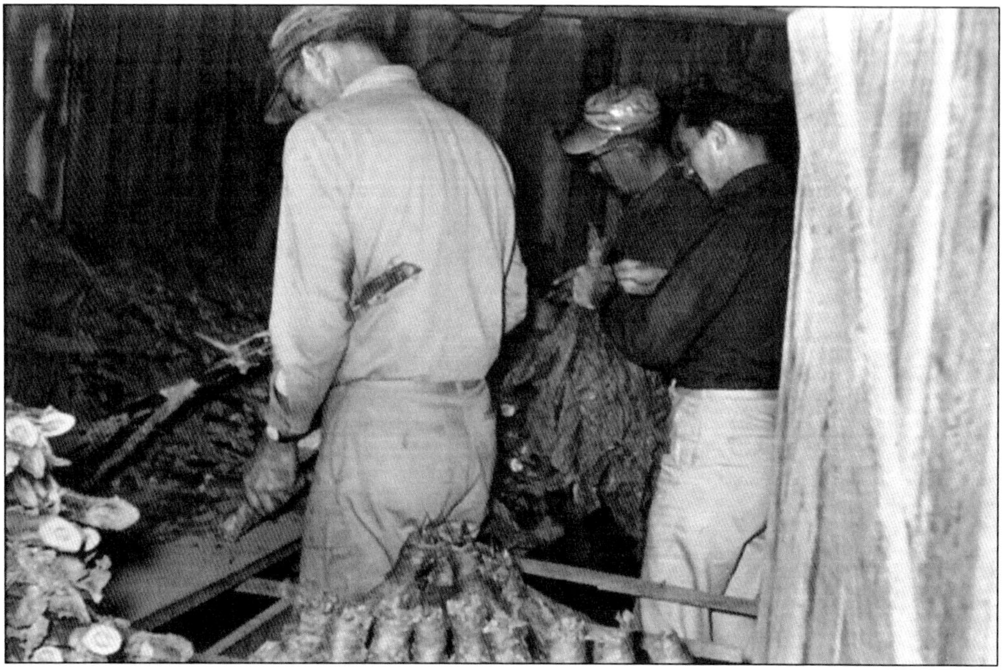

Once the tobacco was cured, it was taken from the barn and into the stripping room. There, the plant was stripped and graded from the stalk and tied into various hands of tobacco. The grades were lugs, tips, reds, and trash. Mike Green (left) and fellow farmers are seen working together in the stripping room. (Courtesy of Green collection.)

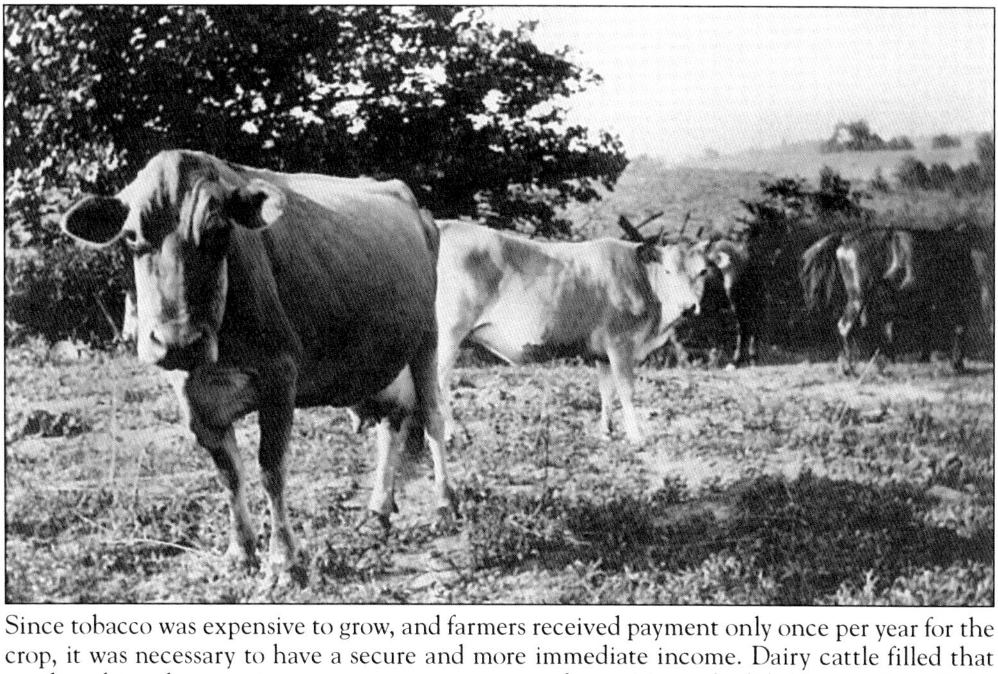

Since tobacco was expensive to grow, and farmers received payment only once per year for the crop, it was necessary to have a secure and more immediate income. Dairy cattle filled that need, making them an important resource on most farms. Not only did they supply milk for sale, but excess milk was also used to make butter and cheese and for cooking. (Courtesy of Bryant collection.)

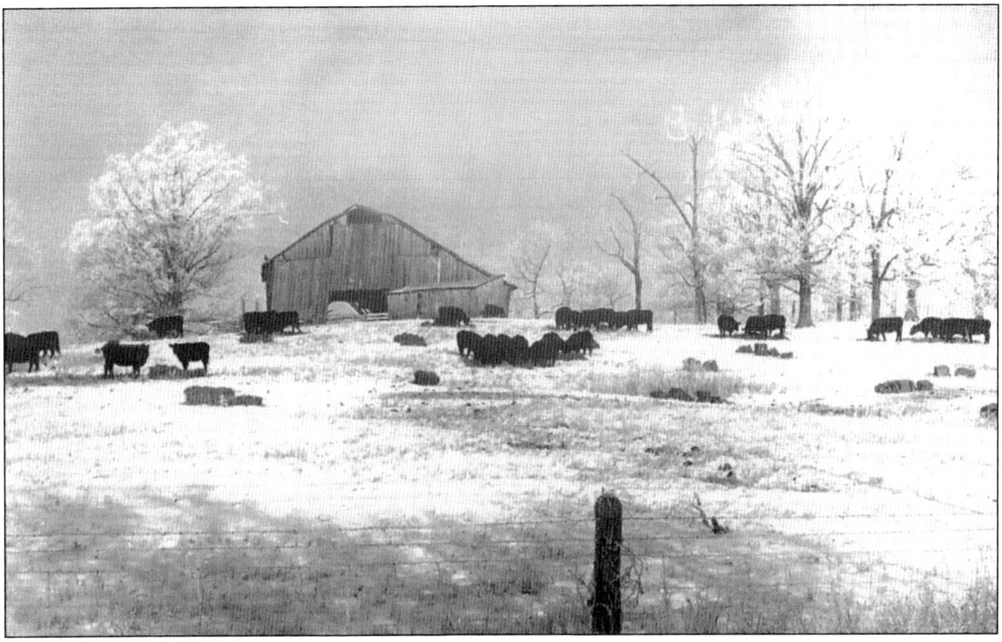

Beef cattle were raised around Lawrenceburg for the meat products they provided, which resulted in good income for farmers. Just like dairy cows, they had to be fed, especially during winter. Farmers provided their animals with bales of hay when the pasture was dead in the winter. (Courtesy of Ripy collection.)

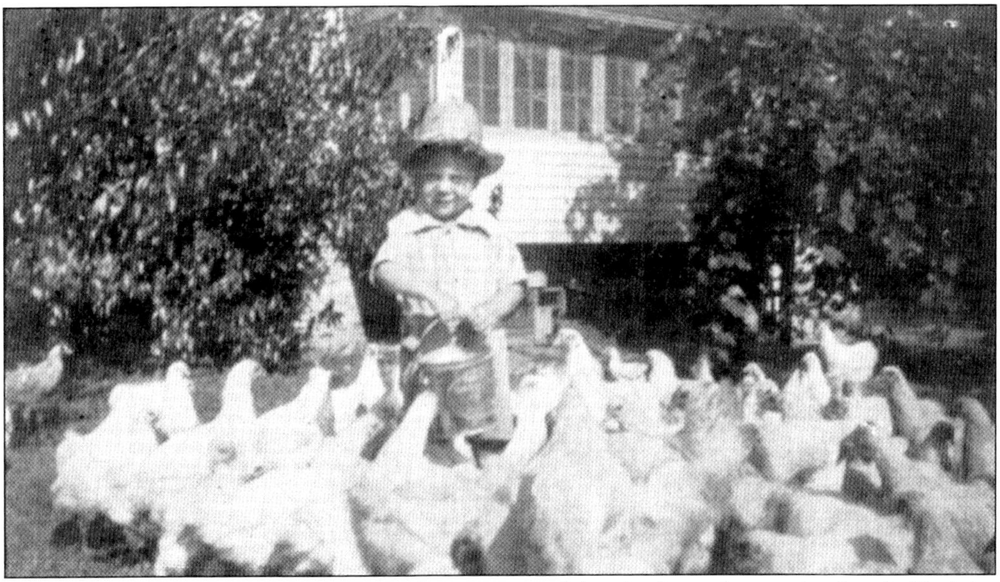

Nearly every yard and lot in Lawrenceburg was used for chickens during the first half of the 20th century. Not only were yards dotted with the birds, chicken coops and pens were seen everywhere. Chickens provided meat and eggs for the table, feathers for pillows and mattresses, and were a ready source of fertilizer. When the preacher came for Sunday dinner, a fat chicken was usually killed, cleaned, and fried for the meal. Around 1929, Willard Bryant's job was to feed and fatten the chickens. Here, he hands them a meal from a bucket filled with cracked corn. (Courtesy of Bryant collection.)

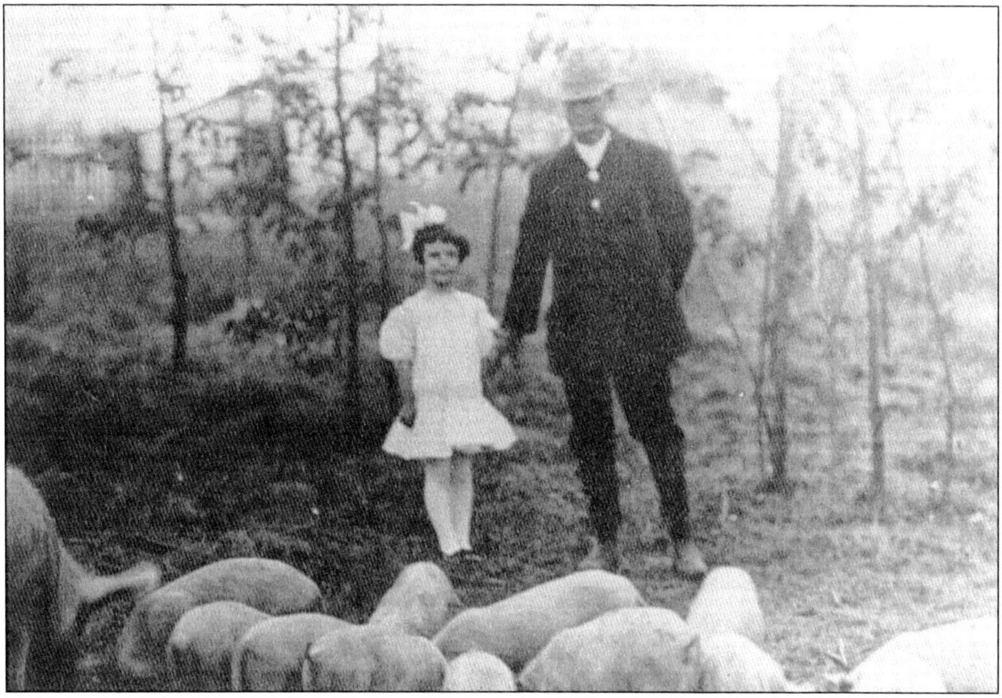

Judge Faris R. Feland walks the grounds of his 30-acre farm on North Main Street with his daughter Catherine Frazier Feland in 1912. Catherine helped him survey a small herd of Tamworth hogs. Swine was raised to expand diets, providing lard, bacon, and other products. The meat was cured before the days of refrigeration. (Courtesy of Feland collection.)

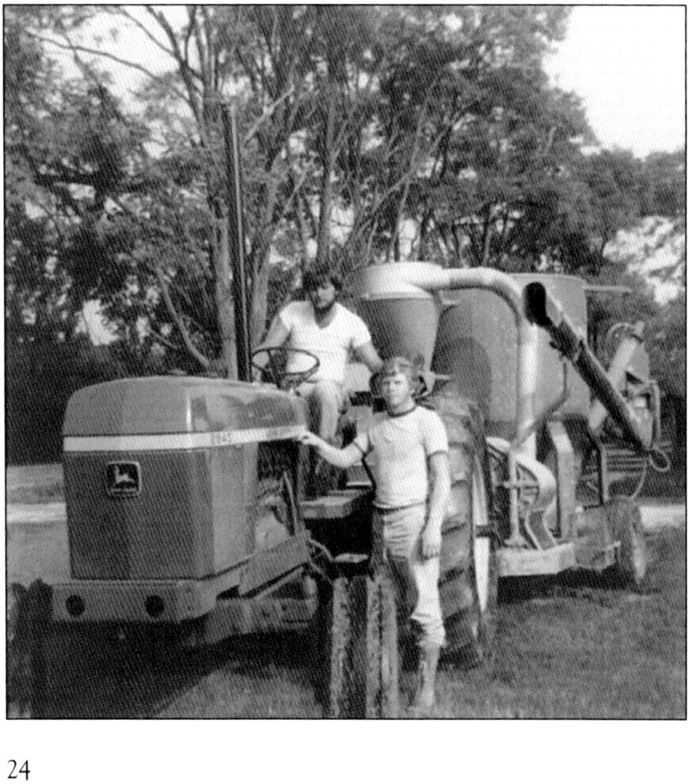

After World War II, tractors and other farm machinery began to replace animals, as they saved on labor costs. The internal-combustion engine required petroleum rather than grass or hay as its fuel source. Here, Mike and Steve Green prepare a tractor and other equipment for use on their farm. (Courtesy of Green collection.)

Three

BOURBON

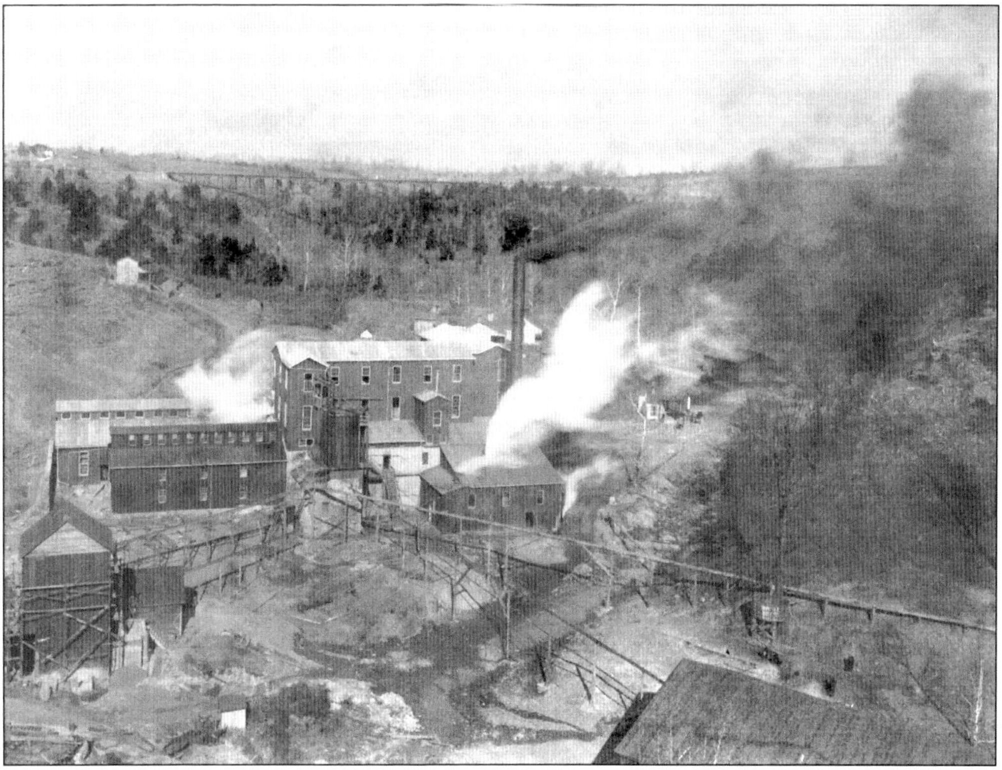

Bourbon has long been associated with Lawrenceburg and Anderson County. Many distilleries have called Lawrenceburg home since 1818, when old Joe Peyton first produced Old Joe. Today, only two distilleries call Lawrenceburg home: Four Roses and Wild Turkey, both of which are part of the Bourbon Trail. Cedar Brook Distillery was established by William H. McBrayer in 1844 and won a number of medals. In 1899, the distillery was sold to Julius Kessler and Company. The water line on the right went to a pump station on the Kentucky River approximately 2.5 miles away. Cedar Brook Distillery was dismantled during Prohibition in 1922. (Courtesy of Anderson County Historical Society.)

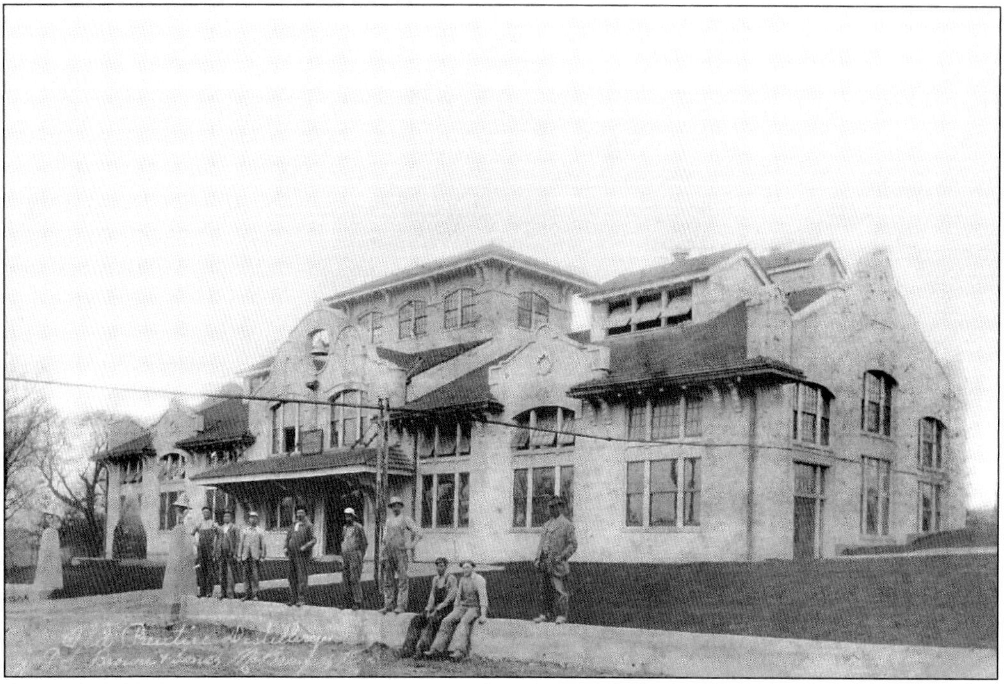

In 1894, J.T.S. Brown and Sons purchased a distillery and changed its name to Old Prentice. In 1910, a new building was constructed on the north side of Bonds Mill Road in a distinctive Spanish mission style. The building, seen here, still stands today and is now known as Four Roses. (Courtesy of Phillips collection.)

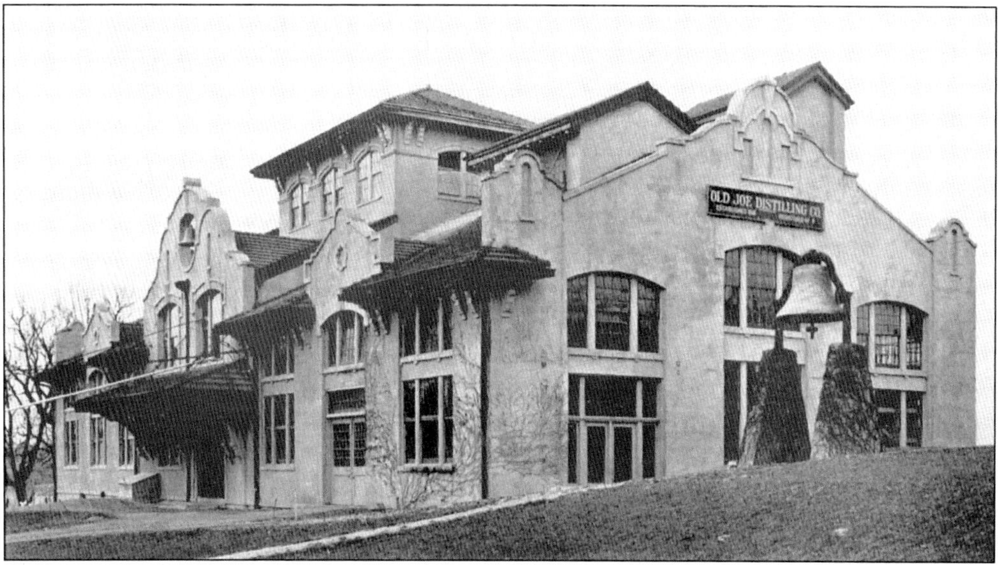

Before the distillery became Four Roses, it operated under a number of different names and owners. During Prohibition, distilling and bottling equipment were removed, and barrel warehouses were dismantled. After the repeal of Prohibition, the distillery was refurbished, new warehouses were built, and it changed ownership again. Old Prentice now became Old Joe Distillery. Since that time, ownership and names have changed, and today the distillery is known as Four Roses. (Courtesy of Anderson County History.)

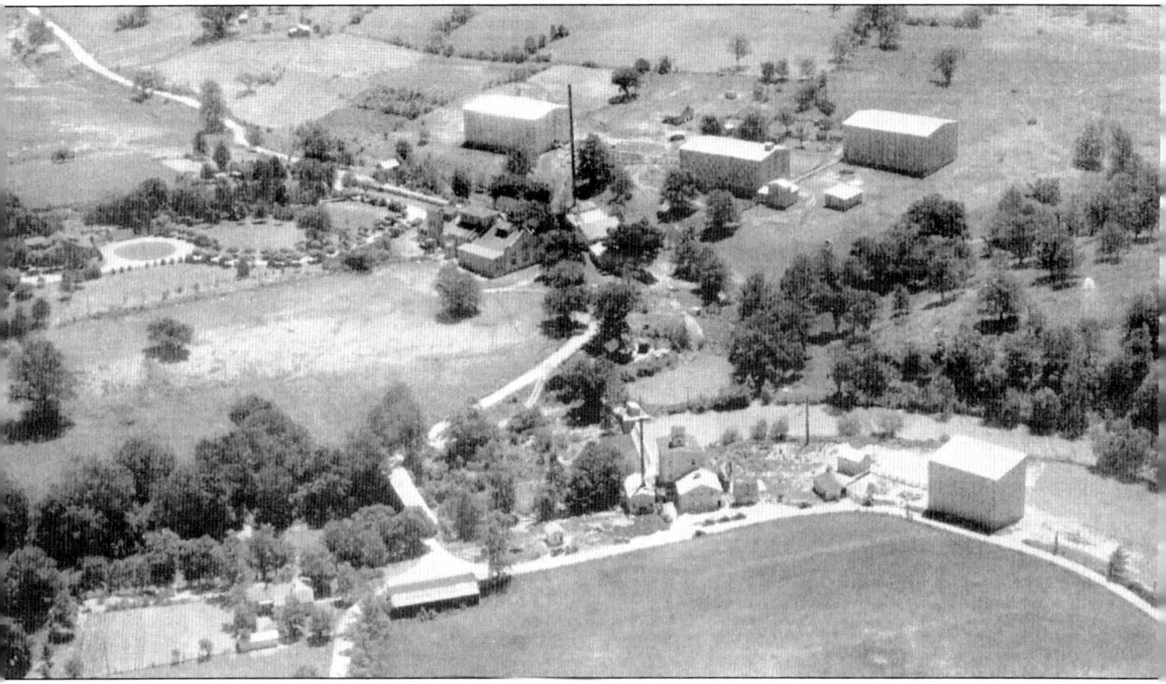

Shortly after Old Prentice became Old Joe and changed ownership, other changes occurred. Montrose, a landscaped estate with a Japanese floral design, was built and occupied by Wilgus Naugher, the secretary-treasurer of the new organization. Montrose and the Old Joe Distillery are seen in the upper half of this photograph. Bonds Mill Distillery is directly across the Salt River. (Courtesy of Anderson County Historical Society.)

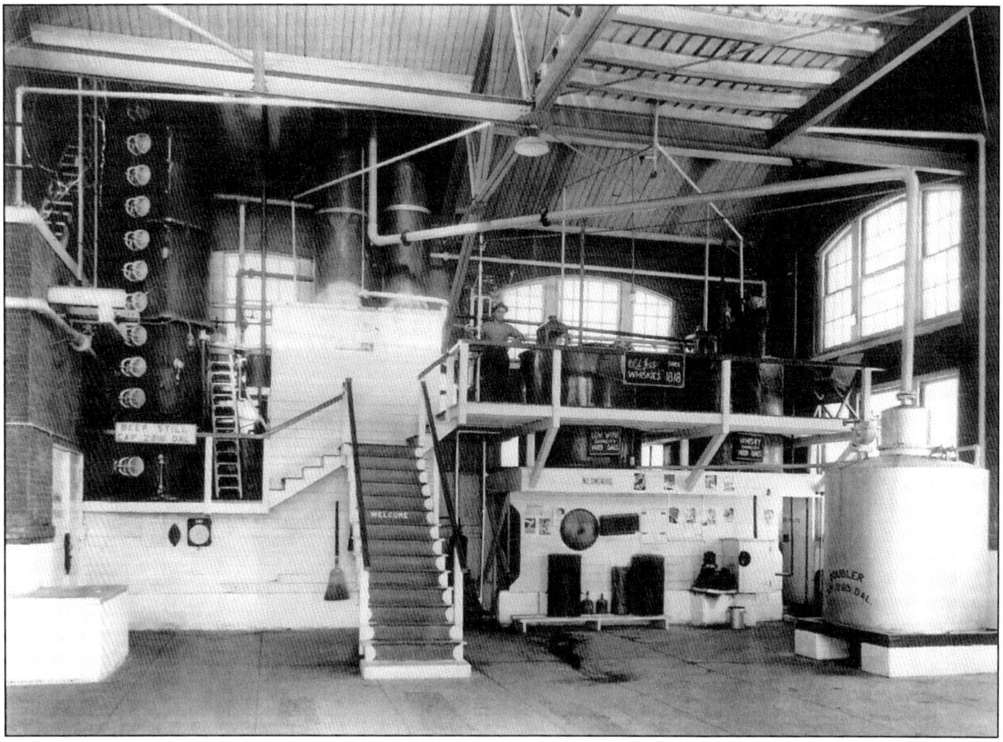

The distilling room of Old Joe shows the beer still, the condensers, and the doubler. (Courtesy of McFarland collection.)

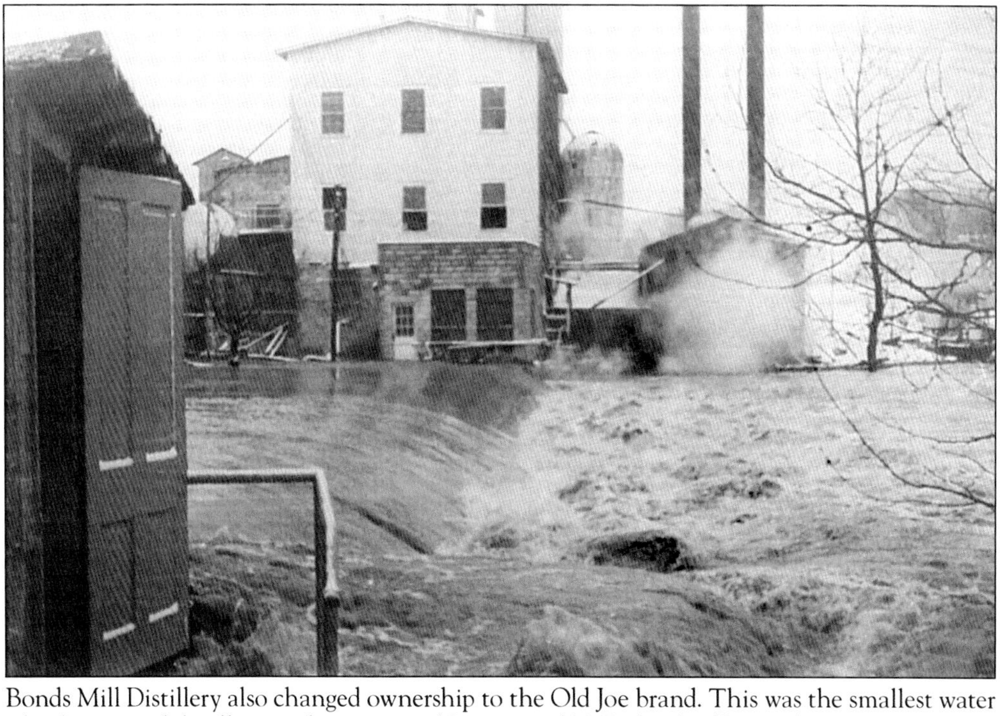

Bonds Mill Distillery also changed ownership to the Old Joe brand. This was the smallest water wheel–operated distillery in the country. (Courtesy of McFarland collection.)

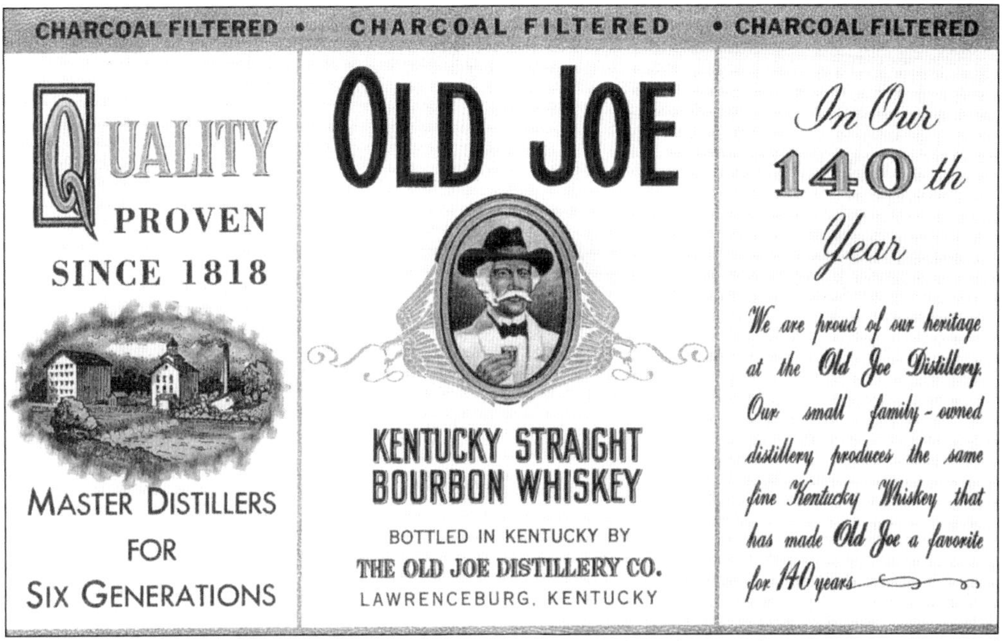

Many historical facts can be gleaned from the Old Joe label, such as its beginnings in 1818 and its six generations of master distillers. It was bottled in Kentucky by the Old Joe Distillery Co., Lawrenceburg, Kentucky, and this particular bottle was celebrating its 140th anniversary. (Courtesy of Wilhoit collection.)

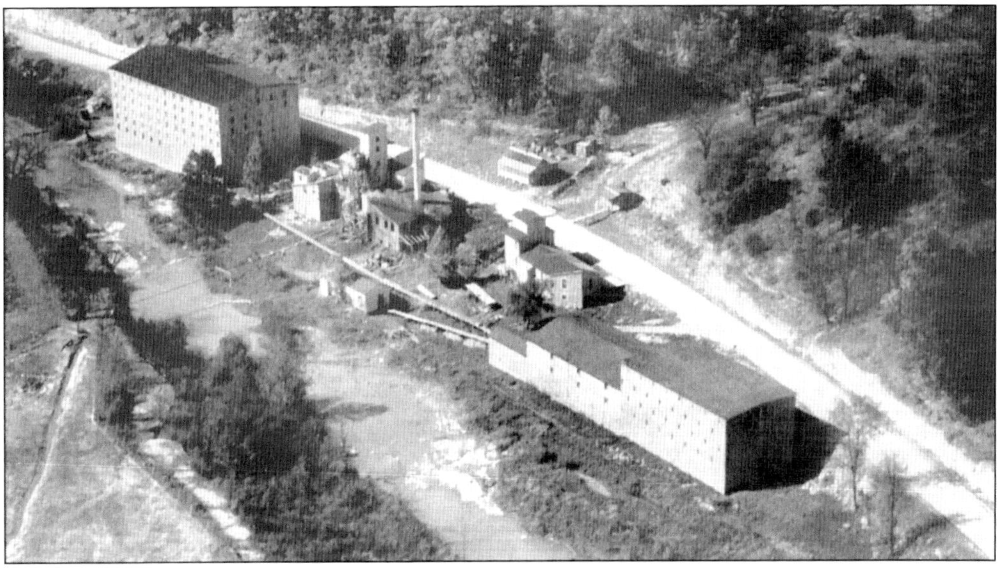

Hoffman Distilling Company sat on the banks of the Salt River six miles west of Lawrenceburg. Although the river was an inconsistent water source for the distillery, never-failing deep wells were drilled to provide water when the river was low. Ezra B. and Robert Ripy owned and operated this distillery, which no longer stands. (Courtesy of Anderson County Historical Society.)

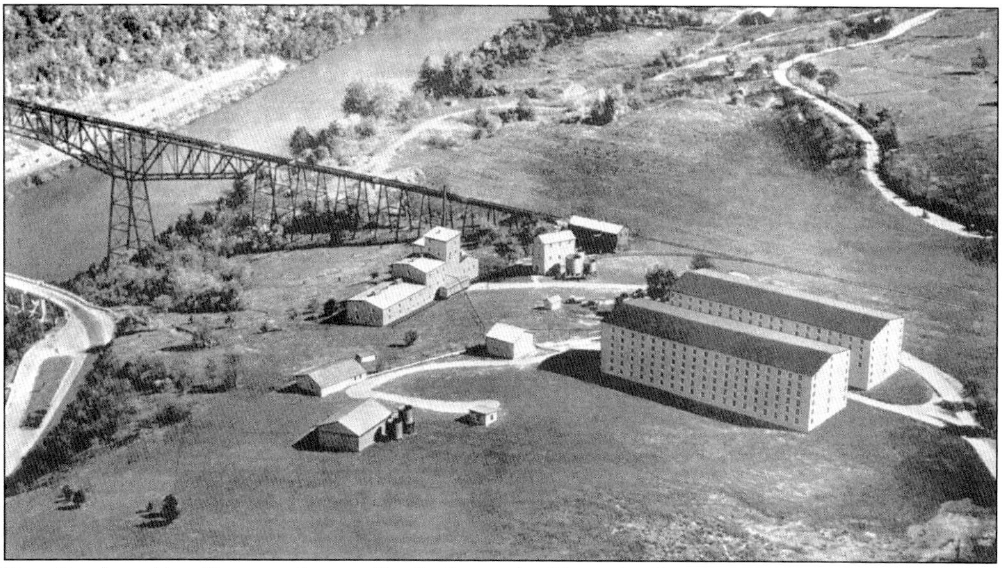

Ripy Brothers Distillery was located high above the Kentucky River near Tyrone. Young's High Bridge and the Blackburn Bridge bordered the grounds of the distillery, which now makes Wild Turkey. (Courtesy of Anderson County Historical Society.)

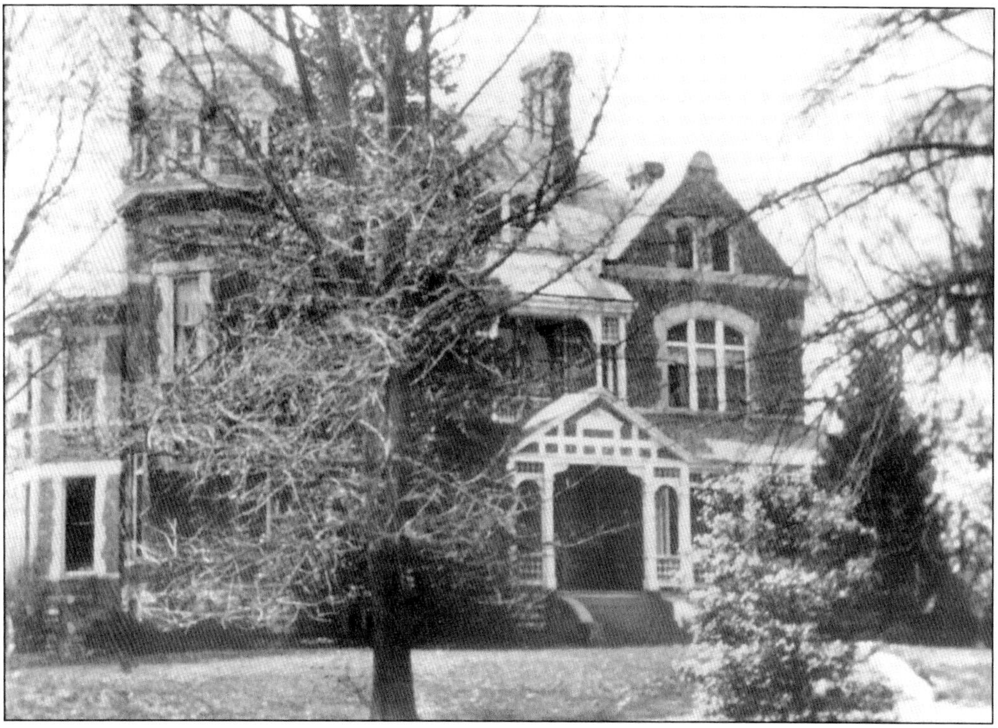

The Thomas B. Ripy home on South Main Street in Lawrenceburg was built in 1885. This elegant mansion passed from the Ripy family and was in ill repair for many years. Members of the family have reacquired the home and are refurbishing and repairing it from years of neglect. (Courtesy of Geoghegan collection.)

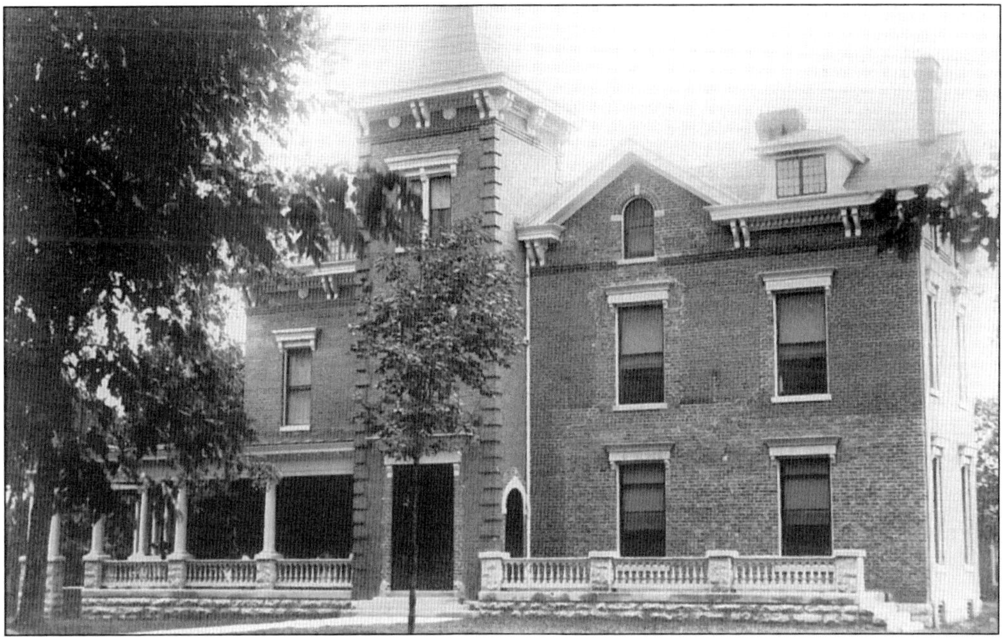

John Dowling came to America from Ireland and settled in Lawrenceburg in 1872. Under his name, three brands of whiskey gained national reputations. He was a partner with Thomas B. Ripy, who then operated Old Dowling. In 1886, Dowling built his home, Dowling Hall, directly across the street from the Ripy mansion. (Courtesy of Hawkins collection.)

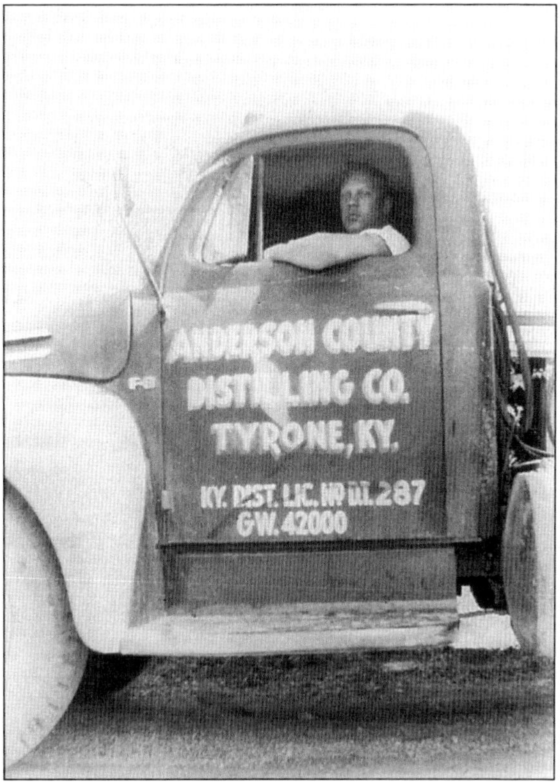

Lawrenceburg was home to many distilleries, and its products were sold far and wide. Every distillery required a dependable transportation system, utilizing trucks and railroads. Early in his career, Willard Bryant was a truck driver for the Anderson County Distilling Company. Not only was whiskey shipped, but raw materials needed for bourbon production were also brought in. (Courtesy of Bryant collection.)

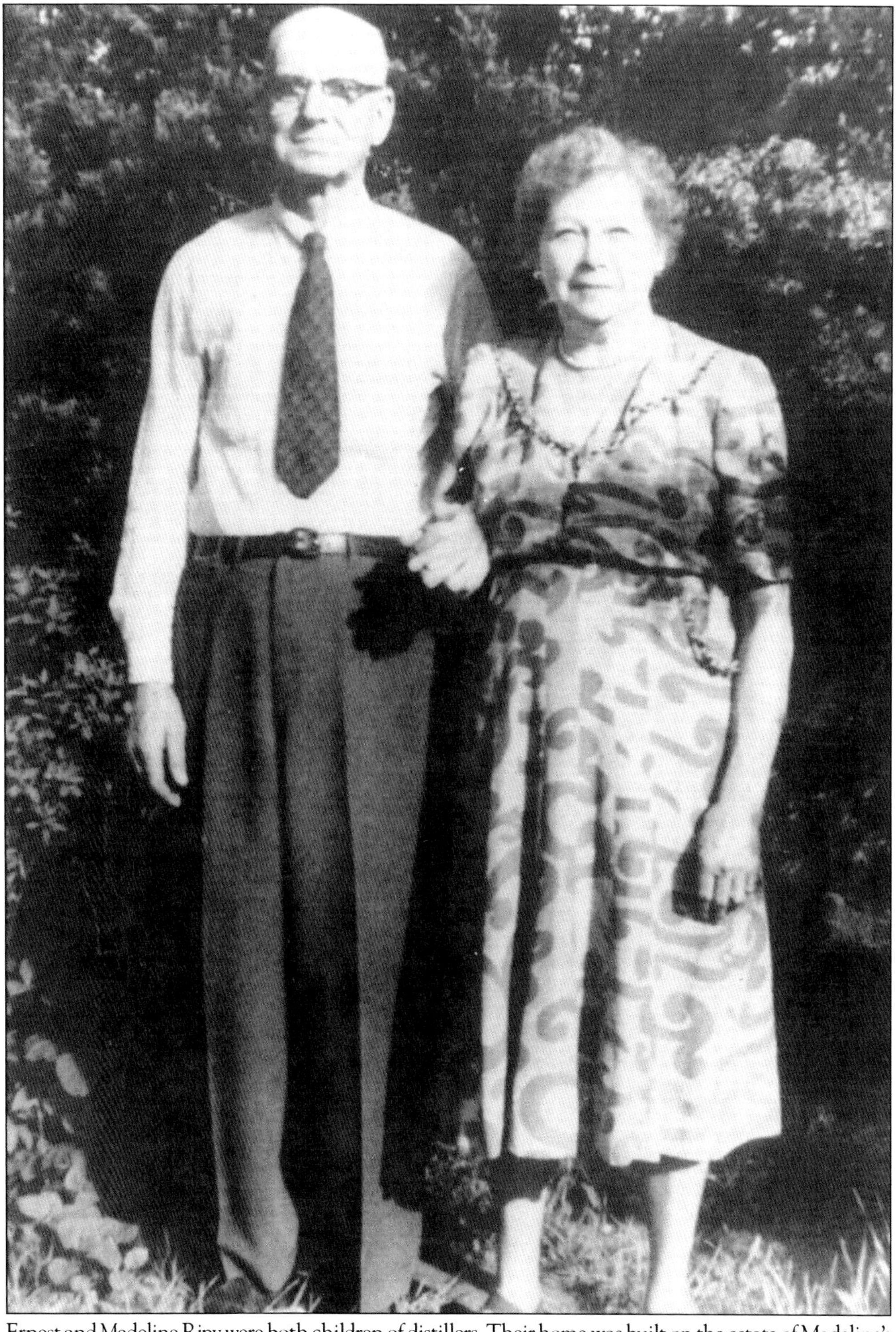

Ernest and Madeline Ripy were both children of distillers. Their home was built on the estate of Madeline's father, Jesse M. Johnson, on South Main Street. (Courtesy of Ripy-Crawford collection.)

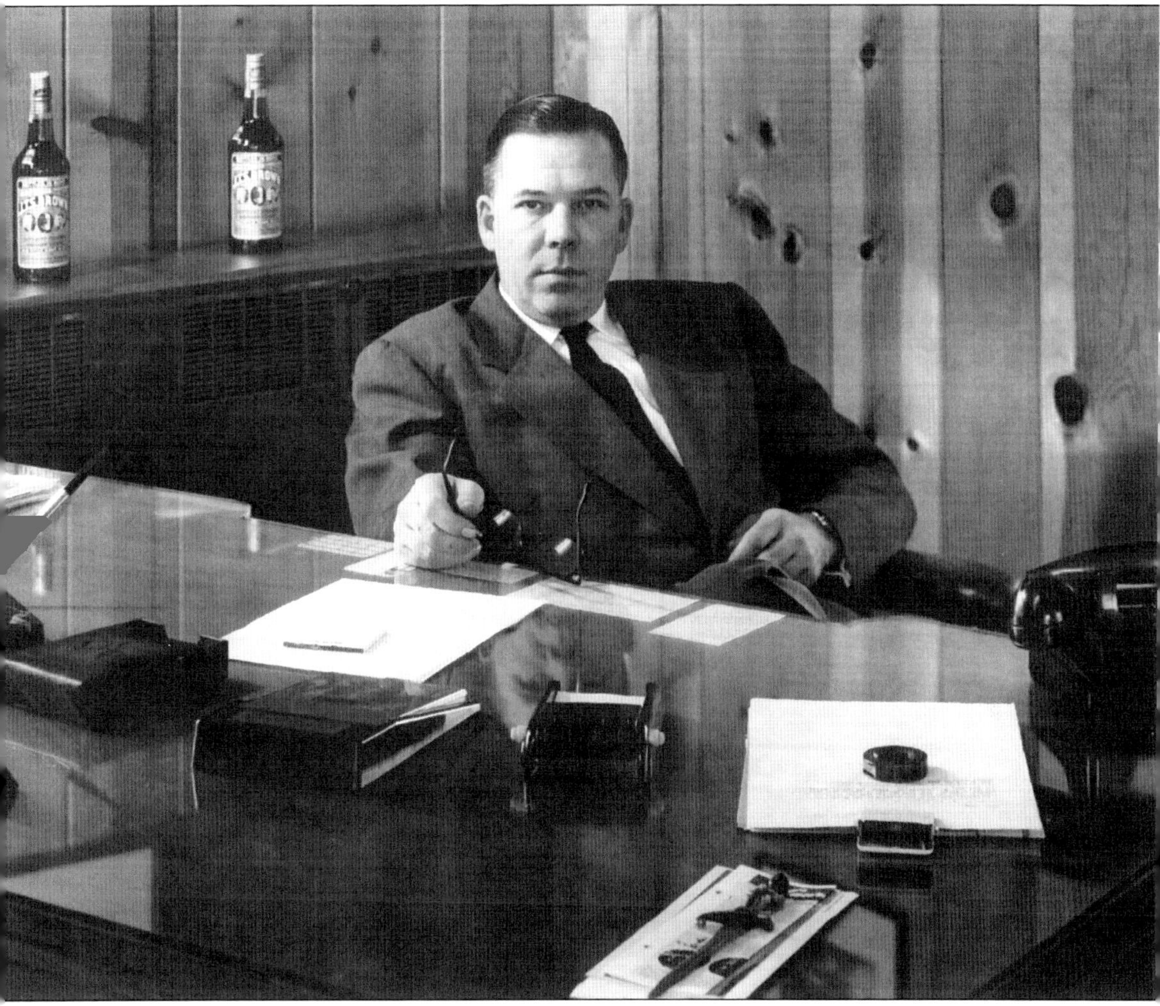

Ernest W. Ripy Jr. was a graduate of Lawrenceburg High School and was the city's first Eagle Scout. He went on to graduate from Purdue University and was a decorated World War II veteran. After the war, he returned to Lawrenceburg, where he assumed a number of roles in the distilling industry. (Courtesy of Ripy-Crawford collection.)

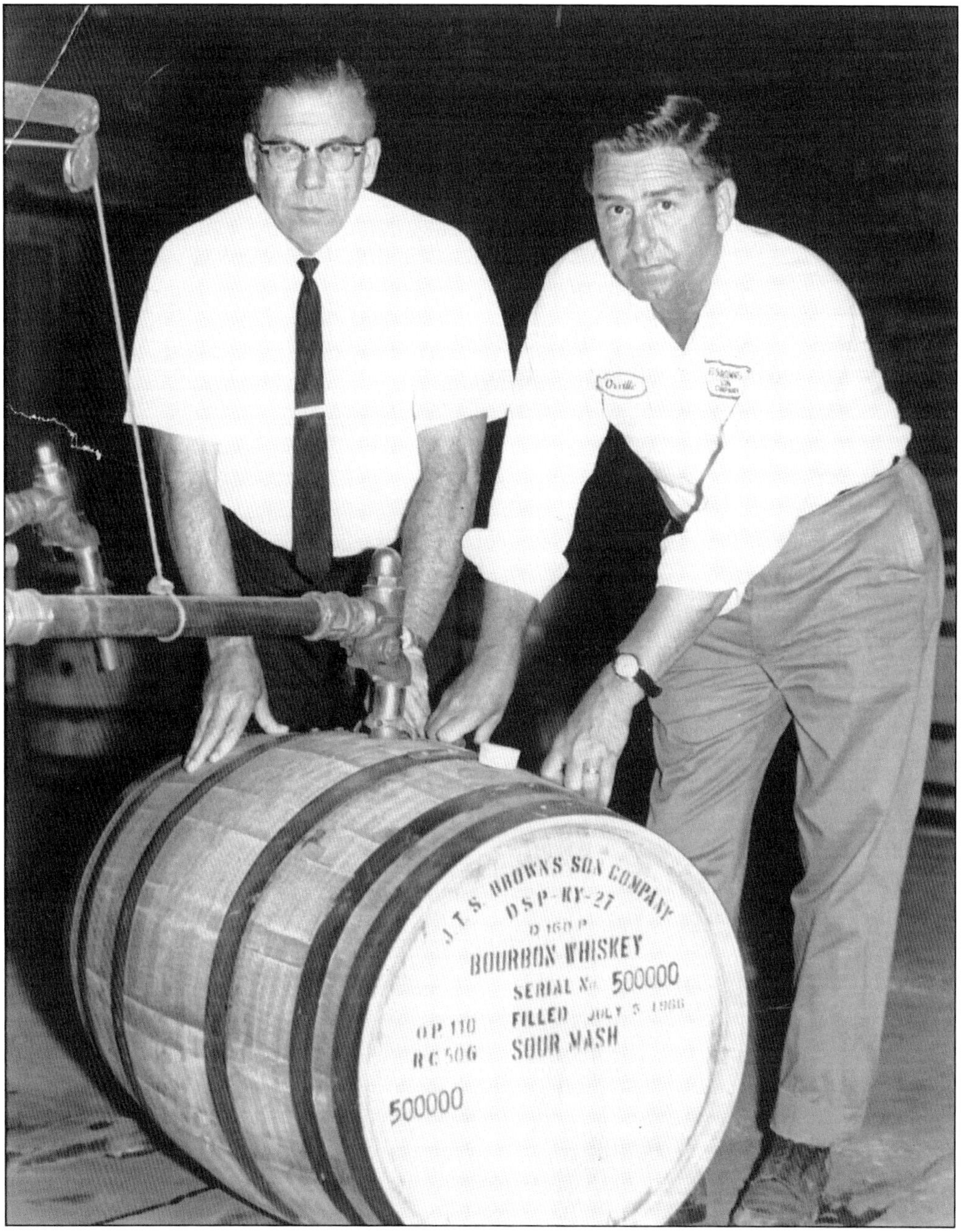

Ernest W. Ripy Jr. (left) and Orville Robinson are seen here with the 400,000th barrel of J.T.S. Brown made in Anderson County. The bourbon was made by the Ripys from 1950 until the late 1970s. Ripy was the president of the Lawrenceburg plant. (Courtesy of Ripy-Crawford collection.)

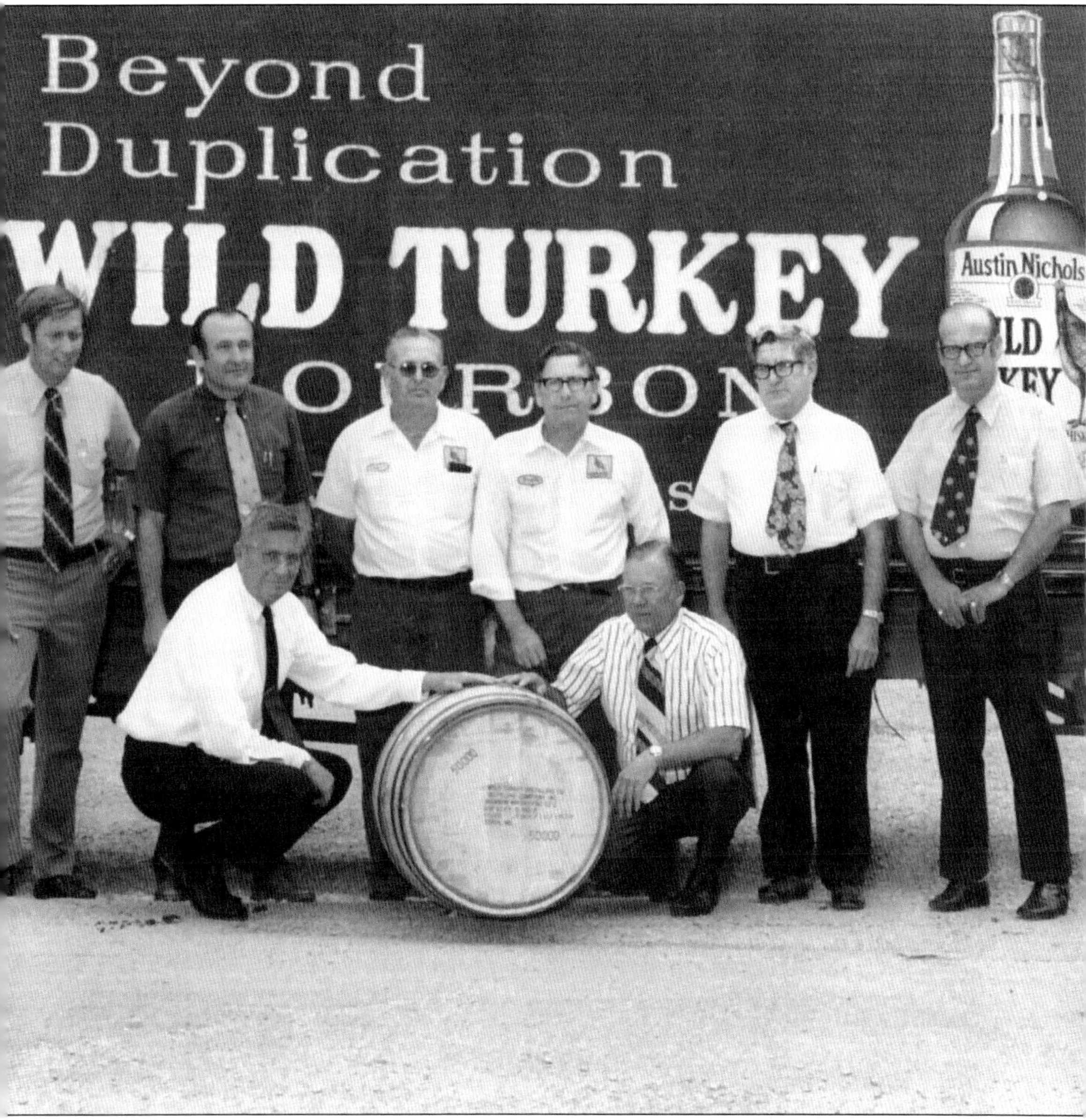

Wild Turkey bourbon replaced J.T.S. Brown in the late 1970s. The company combined employees of the former J.T.S. Brown staff with new members of their own. Jimmy Russel (second from left) became master distiller for Wild Turkey, and Ernest W. Ripy Jr. (right of the barrel) was the vice president. (Courtesy of Ripy-Crawford collection.)

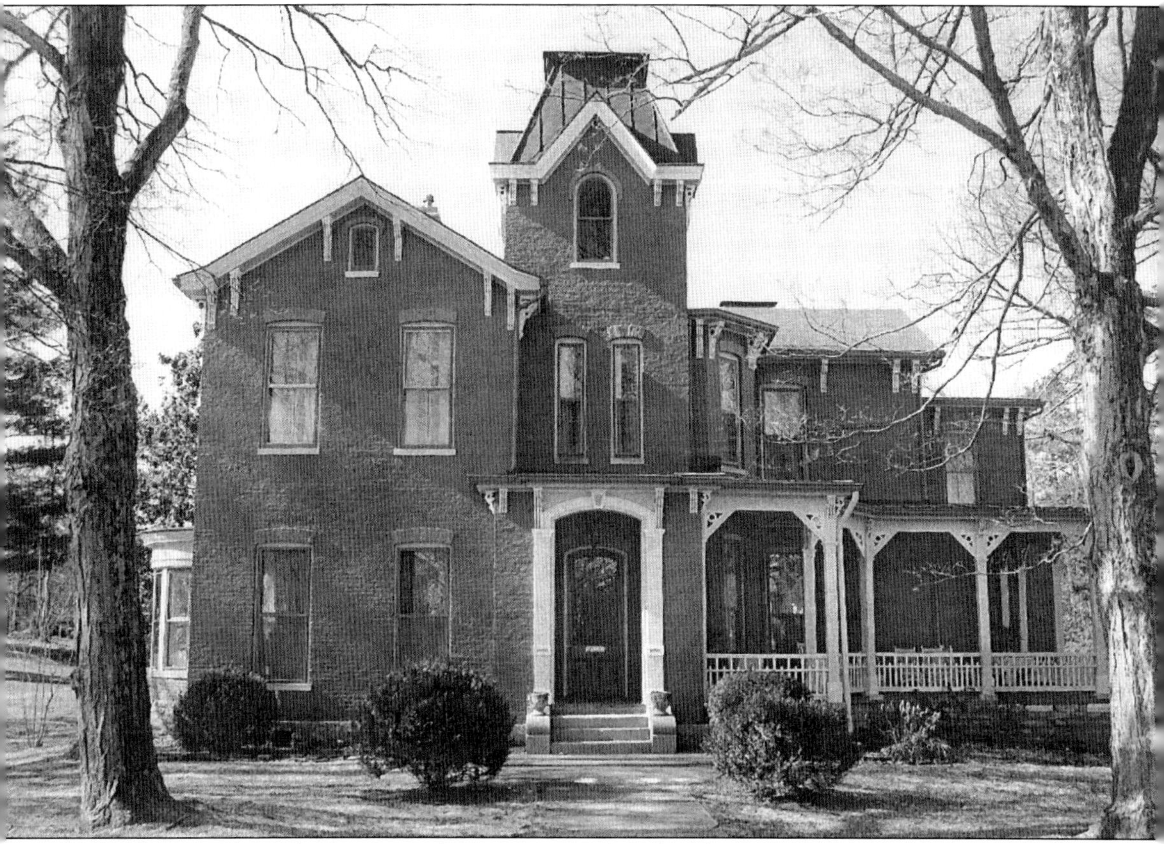

Jesse M. Johnson's home on South Main Street is listed in the National Register of Historic Places. It was built in 1885 by Johnson, who was a banker, a distiller, a horseman, and the first elected mayor of Lawrenceburg. (Courtesy of Wilhoit collection.)

Four

Horse Industry

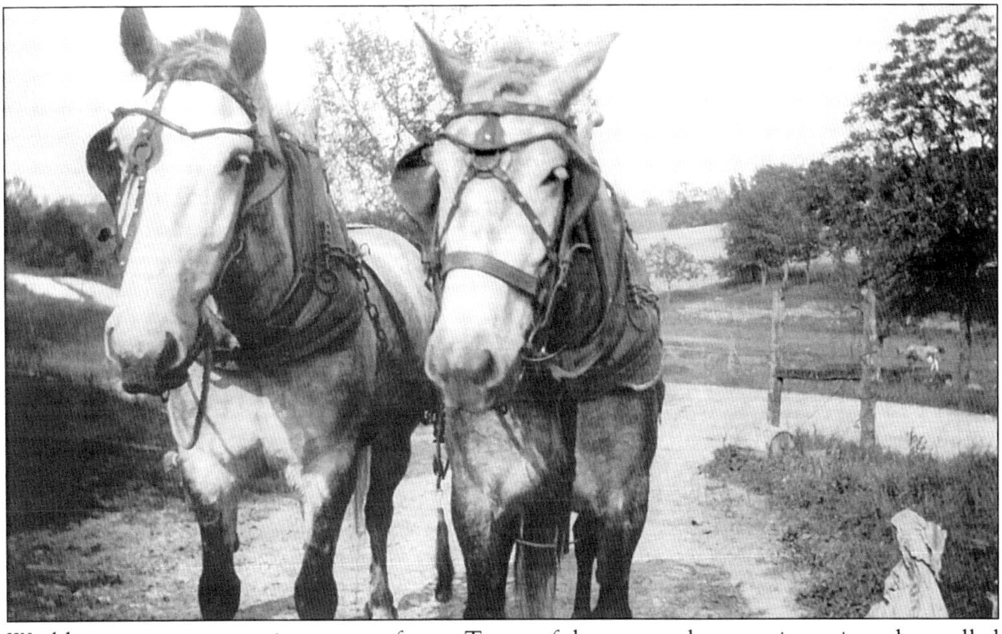

Workhorses were a necessity on most farms. Teams of them were the organic engines that pulled the plow and virtually all other farm equipment. Other industries and occupations revolved around the care and maintenance of the animal. Horses were also used for entertainment, recreation, sport, and other essential services. (Courtesy of Bryant collection.)

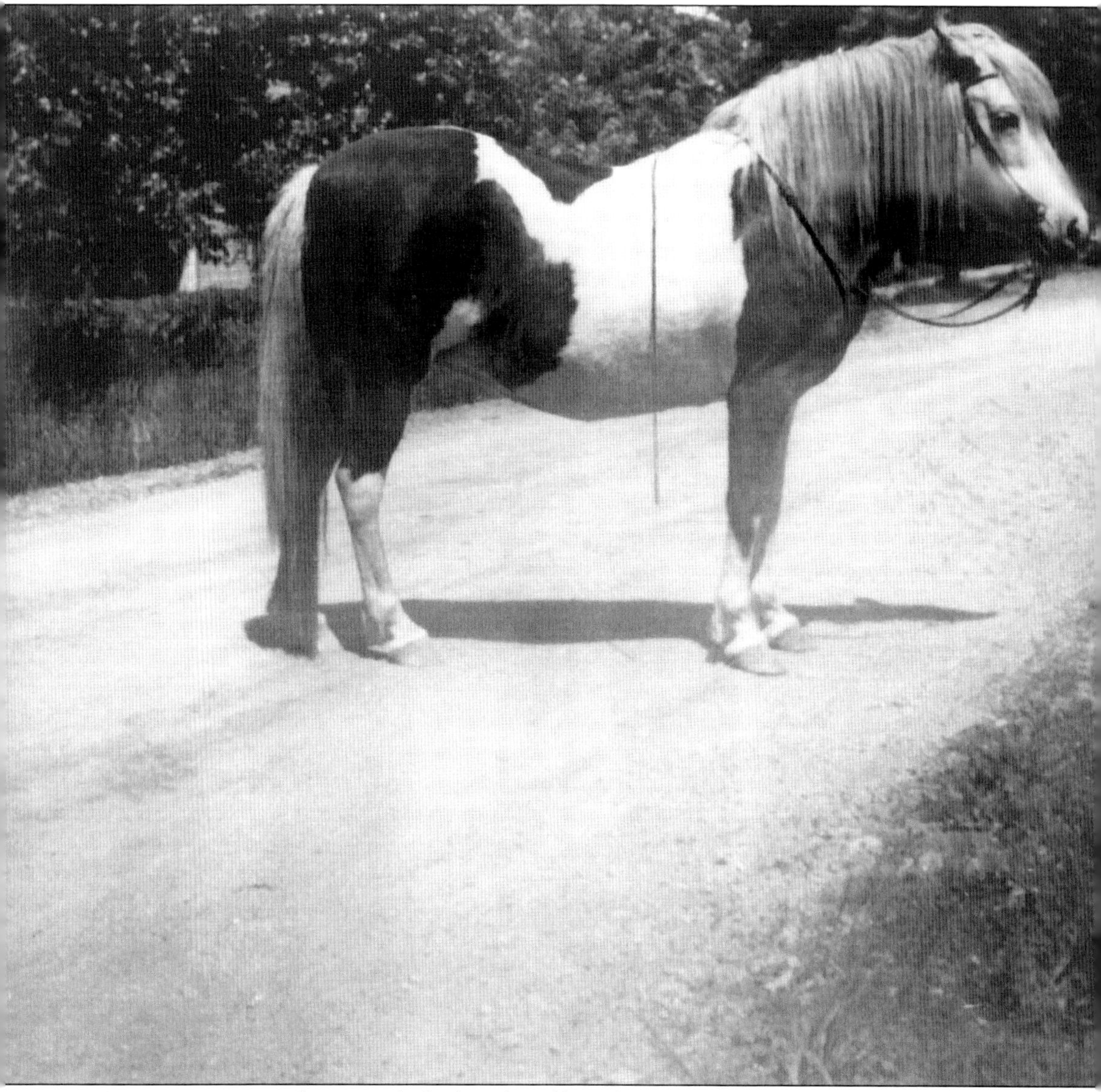

Ponies were commonly seen in the backyards and lots of Lawrenceburg. They provided entertainment for children and adults alike. Jesse M. Johnson imported Shetland ponies to Lawrenceburg from their home in the Shetland Isles. (Courtesy of Bryant collection.)

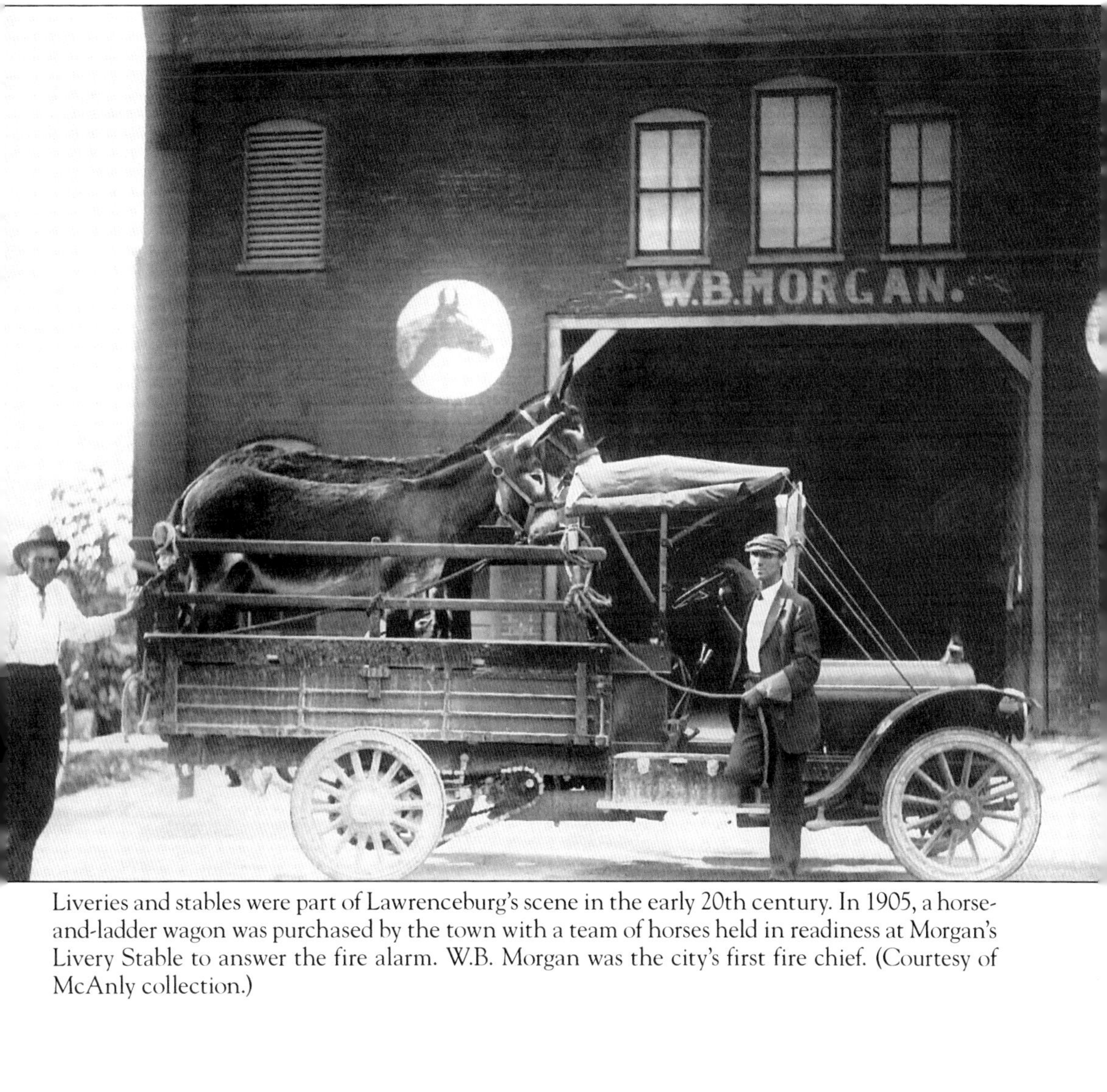

Liveries and stables were part of Lawrenceburg's scene in the early 20th century. In 1905, a horse-and-ladder wagon was purchased by the town with a team of horses held in readiness at Morgan's Livery Stable to answer the fire alarm. W.B. Morgan was the city's first fire chief. (Courtesy of McAnly collection.)

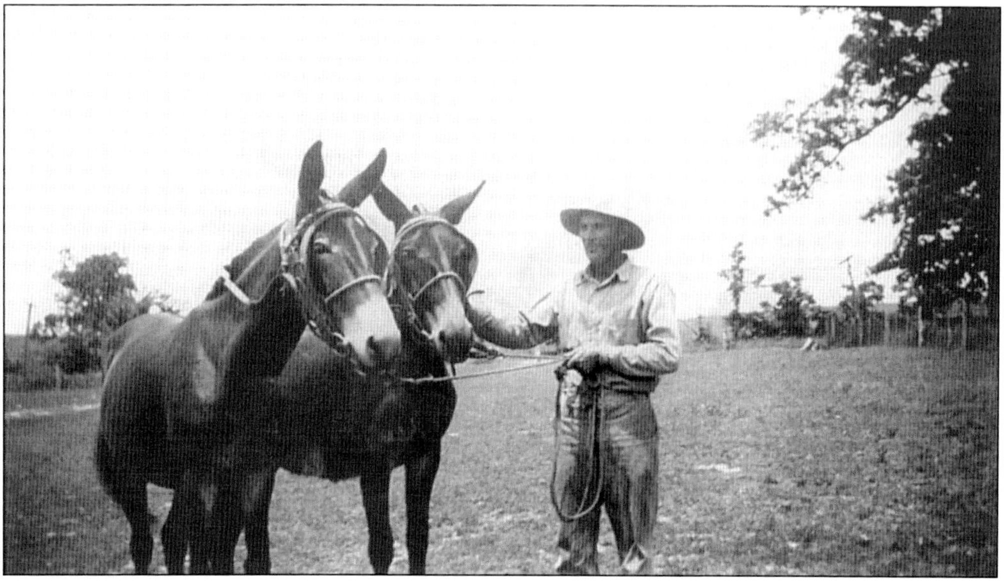

Stanley McMurray is seen here with a team of mules near the village of Ninevah. Kentucky and Anderson County were important producers of mules in the 19th century. (Courtesy of Bryant collection.)

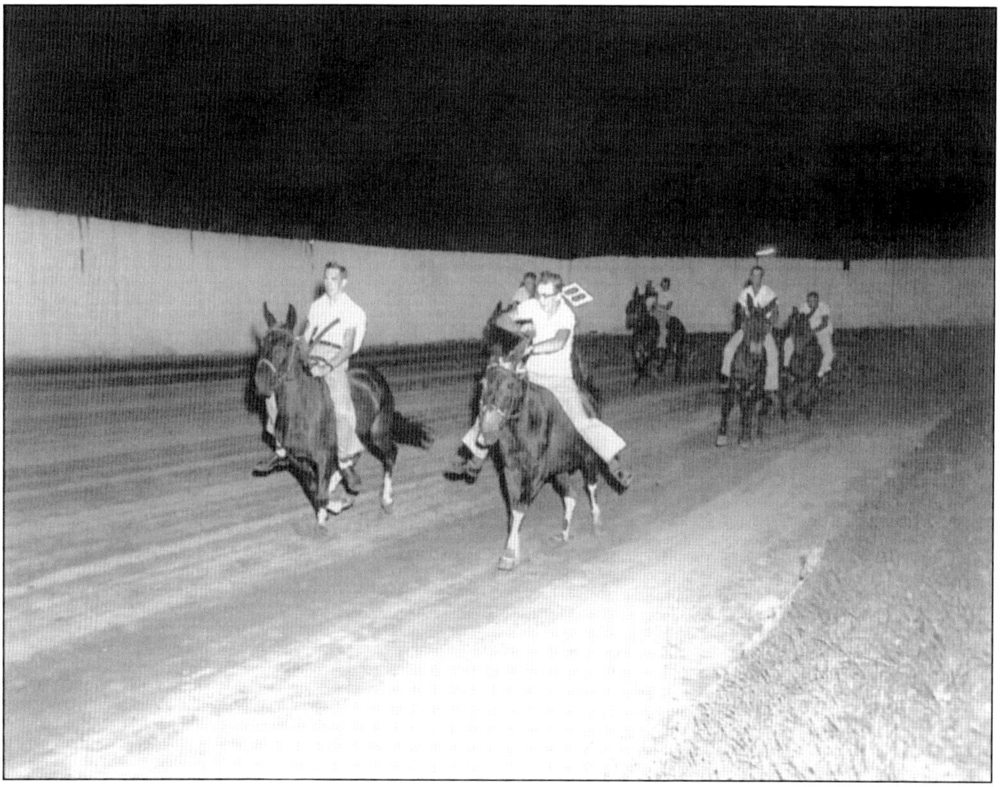

Mule races were always an enjoyable event at the Lawrenceburg Fair and Horse Show. Here, Jackie Sutherland leads a close race at the fair in the 1950s. (Courtesy of Hanks collection.)

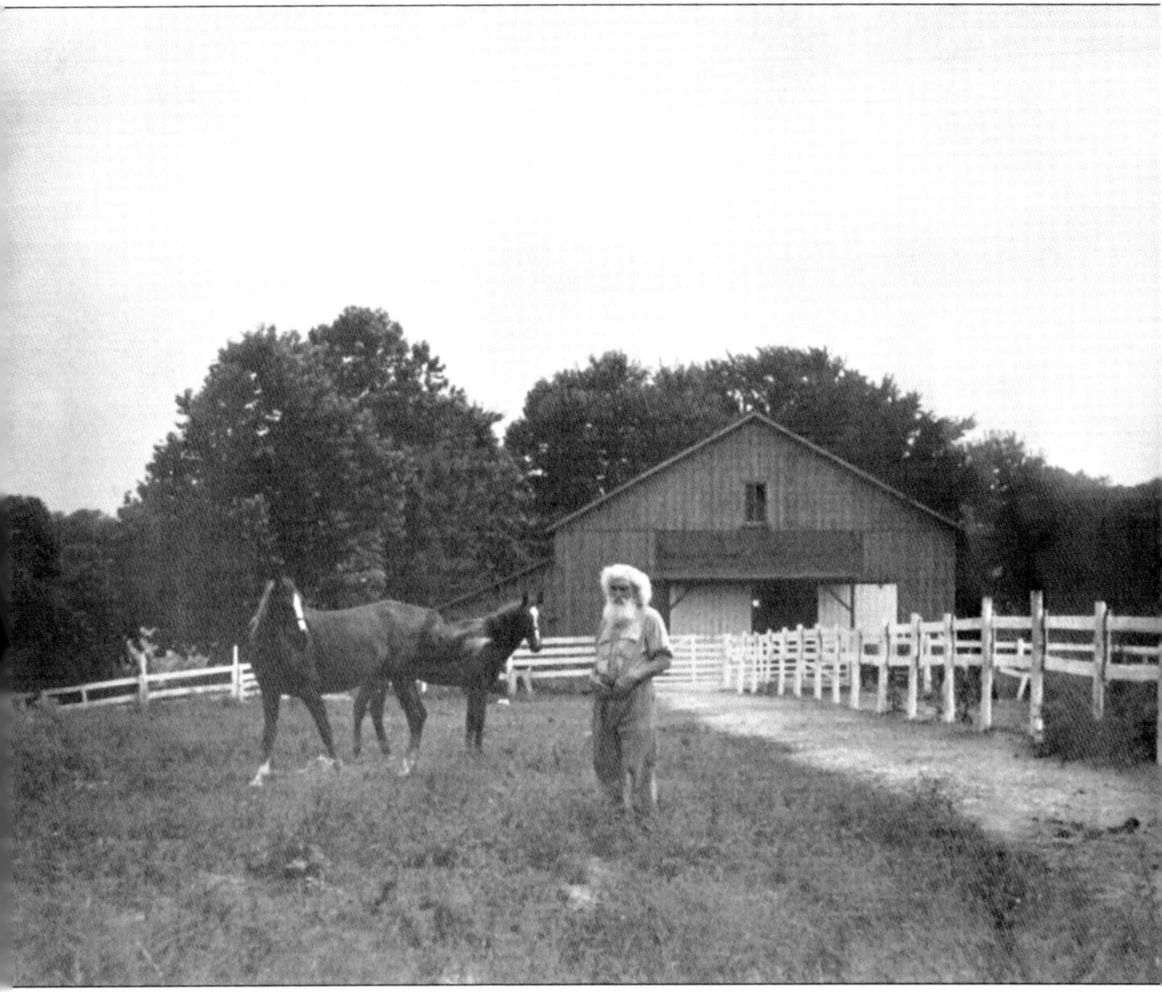

Ben Wilson was renowned for his knowledge of good horses. He is seen here giving advice to two attentive saddle horses in Anderson County. Ben traveled—mostly on foot—to horse events across the state, including the state fair in Louisville. (Courtesy of Smith collection.)

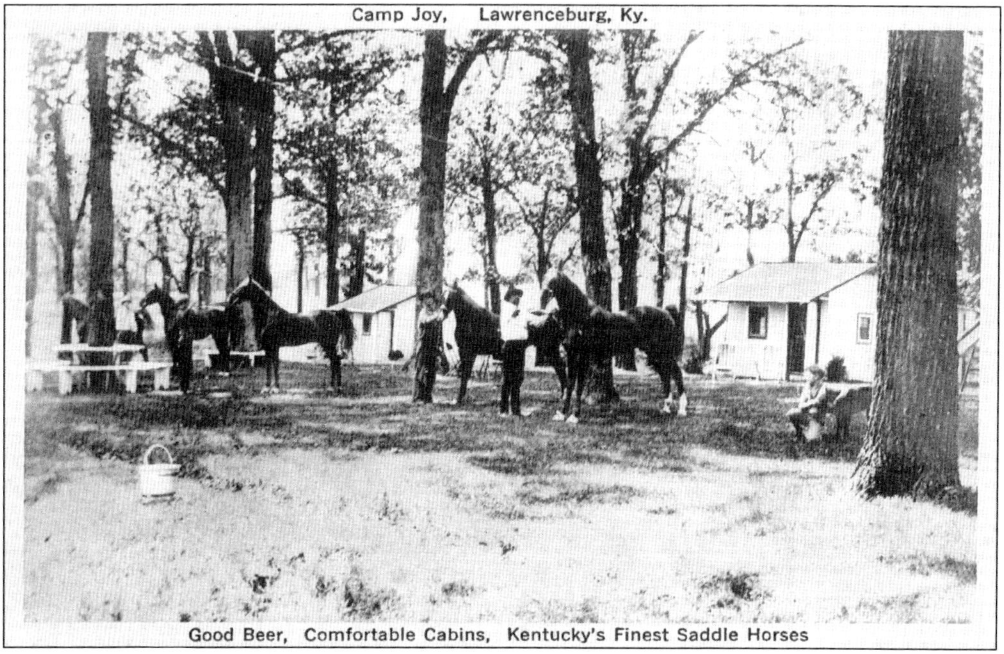

Camp Joy, located off Highway 62, advertised itself as a place for good food, comfortable cabins, and Kentucky's finest saddle horses to travelers across the state. The cabins and horses are seen here. (Courtesy of McAnly collection.)

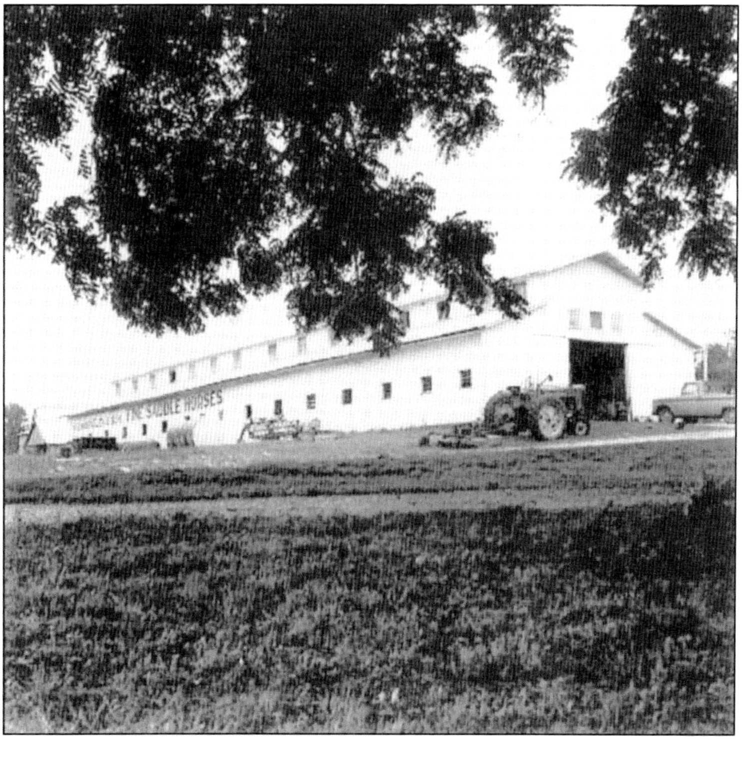

This Mountjoy Fine Saddle Horses Barn sat on the Walnut Grove Farm off Highway 62. This farm was the dealer and handler of high-class horses, especially walk and trot horses. Notable horses included Judge Collins, Jack Starbright, Lady Worthington, and Maid of Honor, among others. (Courtesy of Mountjoy collection.)

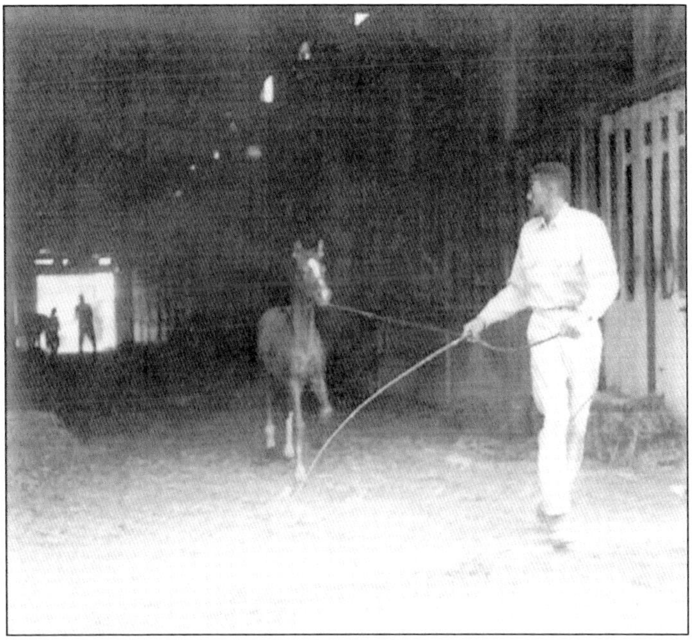

Walter Johnson trains a colt on the Mountjoy farm. The training of horses began early in the life of a colt and continued through its adulthood. (Courtesy of Mountjoy collection.)

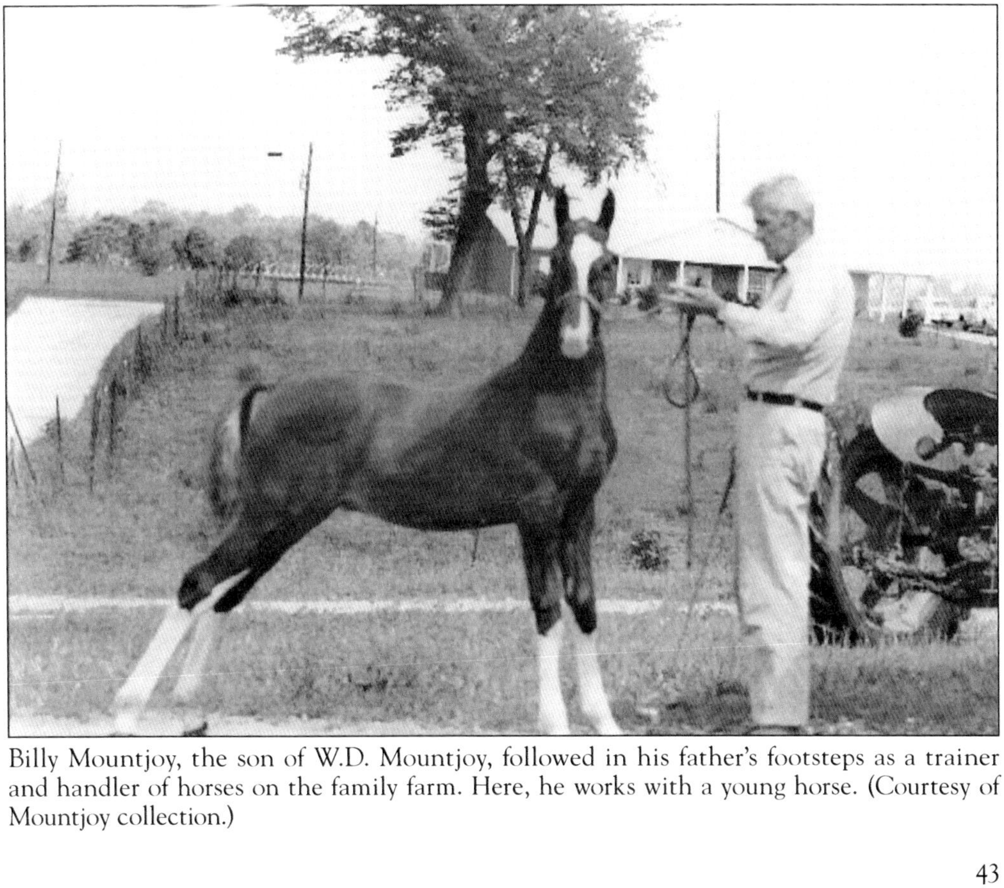

Billy Mountjoy, the son of W.D. Mountjoy, followed in his father's footsteps as a trainer and handler of horses on the family farm. Here, he works with a young horse. (Courtesy of Mountjoy collection.)

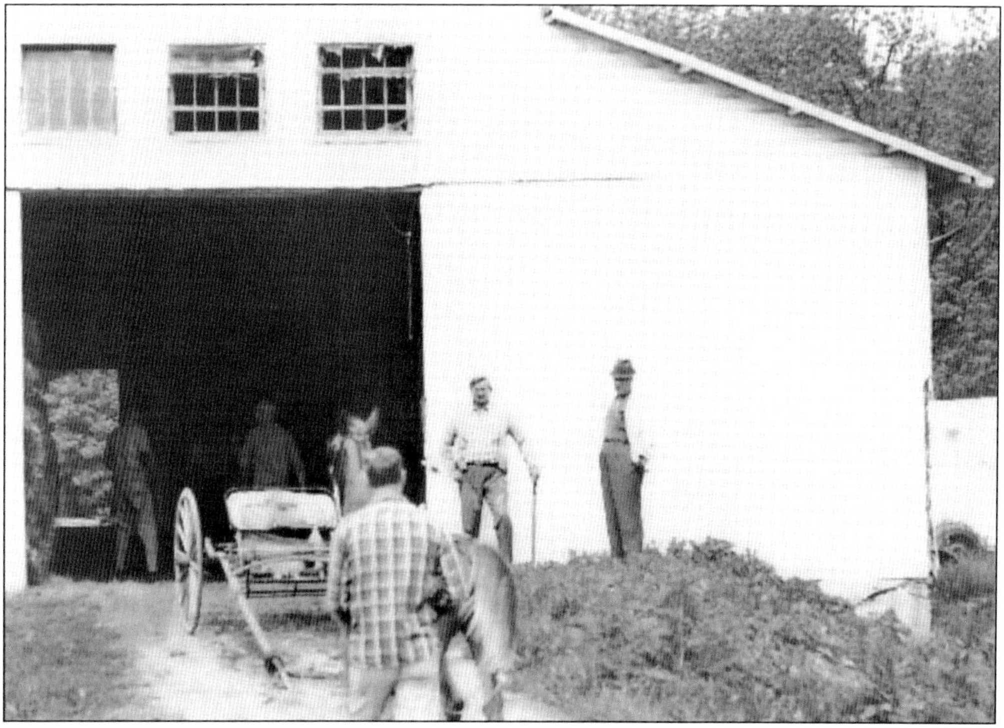

Jimmy Mountjoy (center), the grandson of W.D. Mountjoy, watches as another trainer leads a horse to its daily exercise. (Courtesy of Mountjoy collection.)

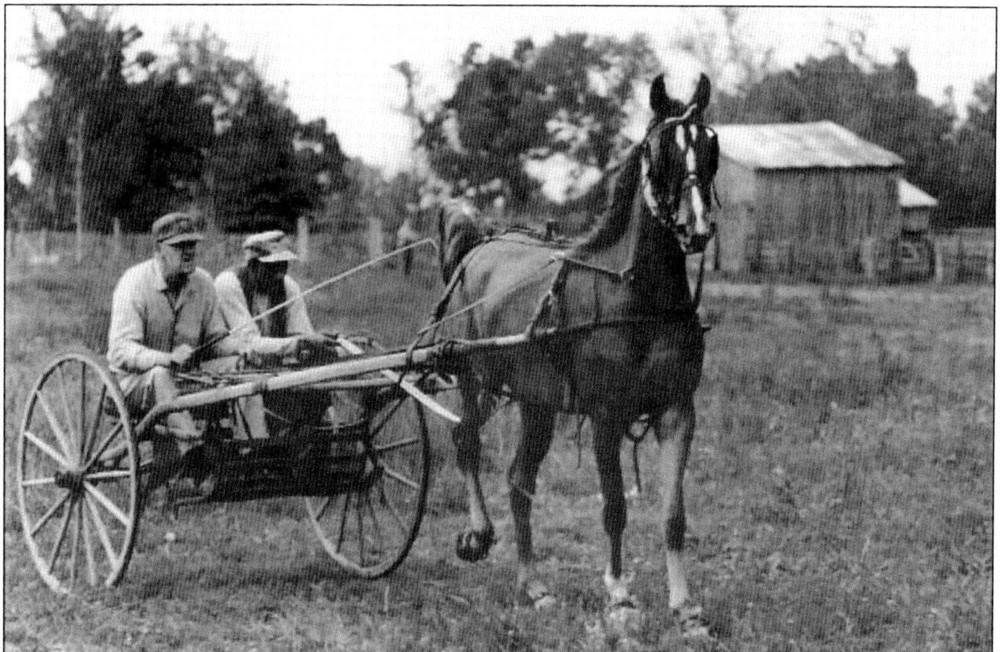

Billy Mountjoy and Walter Johnson put a trotter through its paces during a training exercise on the Mountjoy horse farm. The training of horses was an intensive, hands-on occupation, and good trainers got the best from their animals. (Courtesy of Mountjoy collection.)

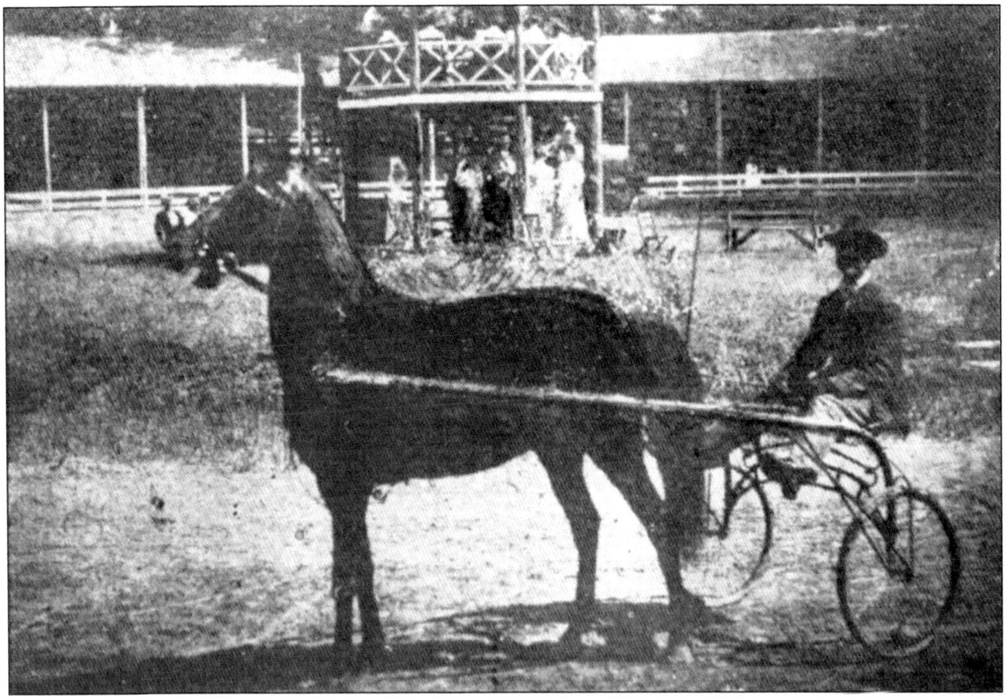

W.D. Mountjoy is seen here at the Lawrenceburg Fair and Horse Show in 1900 just after this two-year-old gelding owned by Eli Shelburne won a blue ribbon. Mountjoy was a Hall of Fame horse trainer. (Courtesy of Hawkins collection.)

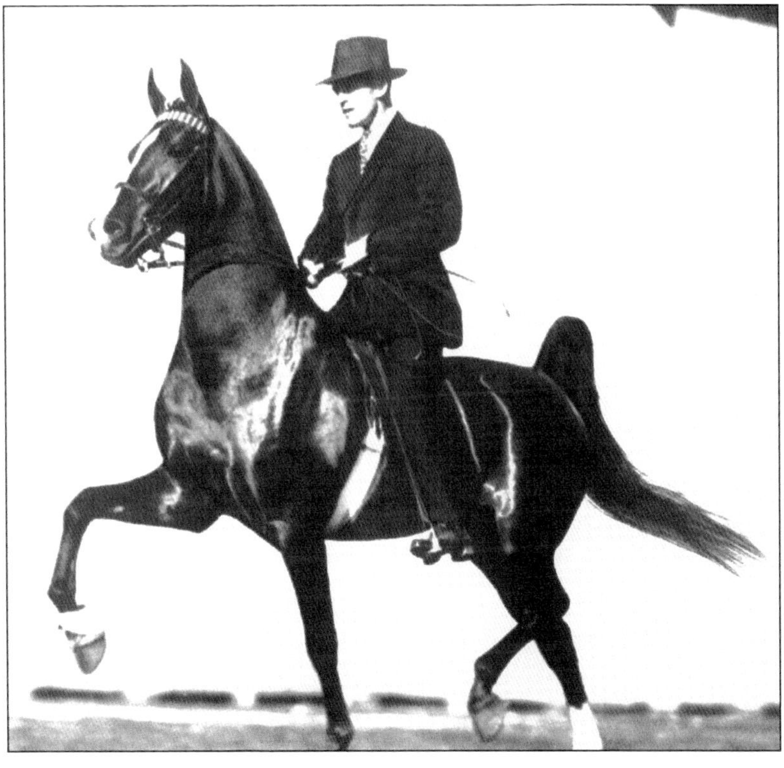

Miller McAfee and other members of his family spent their lives training, showing, and riding horses. McAfee was a world-class rider and later became a renowned equestrian judge. (Courtesy of Major collection.)

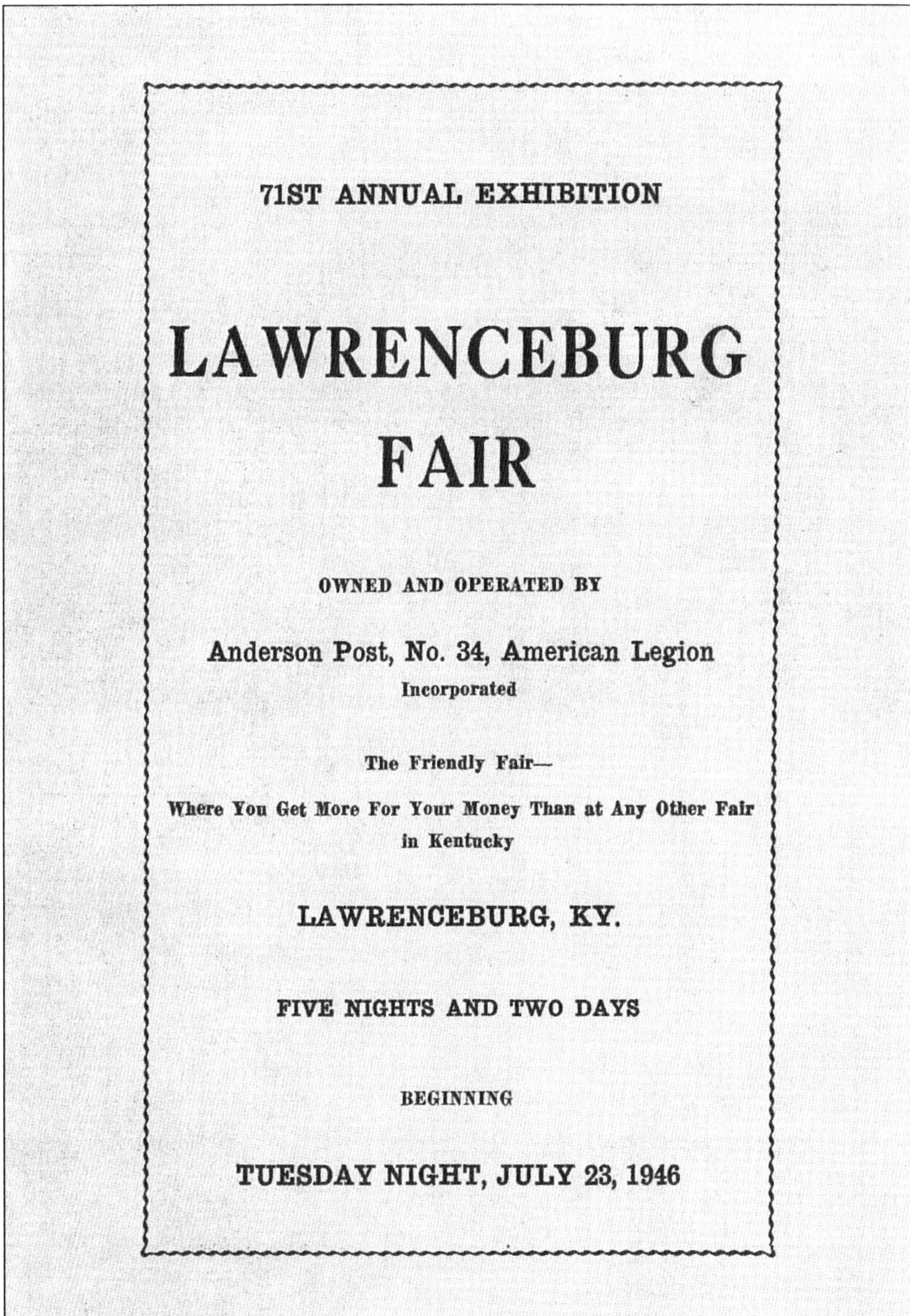

71ST ANNUAL EXHIBITION

LAWRENCEBURG FAIR

OWNED AND OPERATED BY

Anderson Post, No. 34, American Legion

Incorporated

The Friendly Fair—

Where You Get More For Your Money Than at Any Other Fair in Kentucky

LAWRENCEBURG, KY.

FIVE NIGHTS AND TWO DAYS

BEGINNING

TUESDAY NIGHT, JULY 23, 1946

The Lawrenceburg Fair and Horse Show was owned and operated by the American Legion Anderson Post No. 34. This brochure was for the annual exhibition on July 23, 1946. The American Legion continues to operate the fair. (Courtesy of Hawkins collection.)

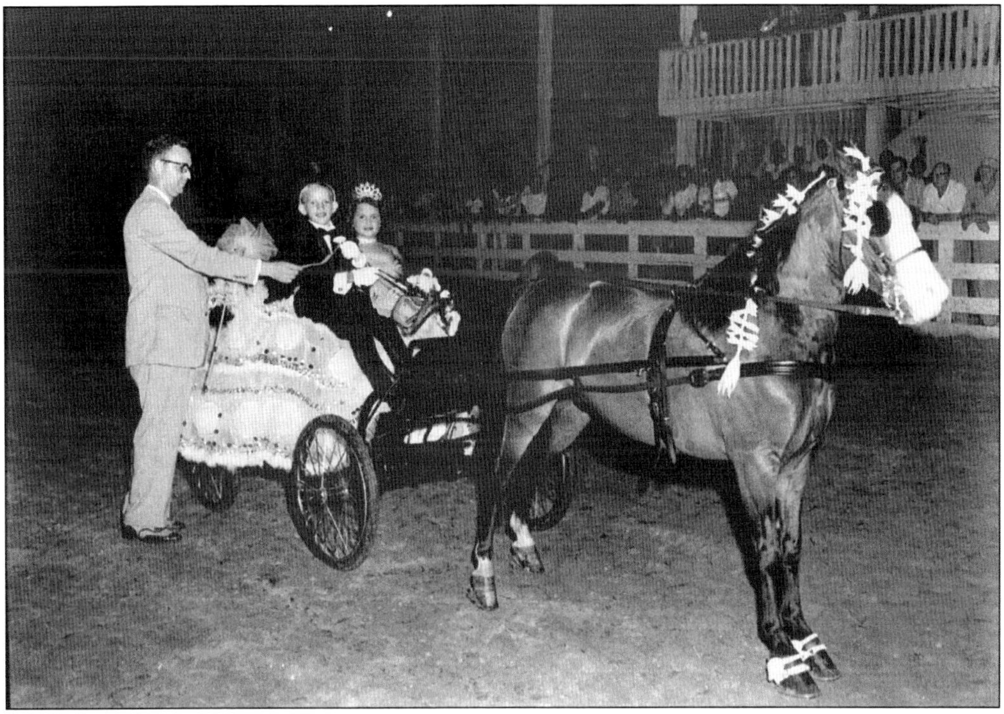

Walter Patrick served for many years and wore many hats at the Lawrenceburg Fair and Horse Show. He was an announcer, a ringmaster, and the legion post commander. He is seen here presenting a ribbon at the fair. (Courtesy of Hanks collection.)

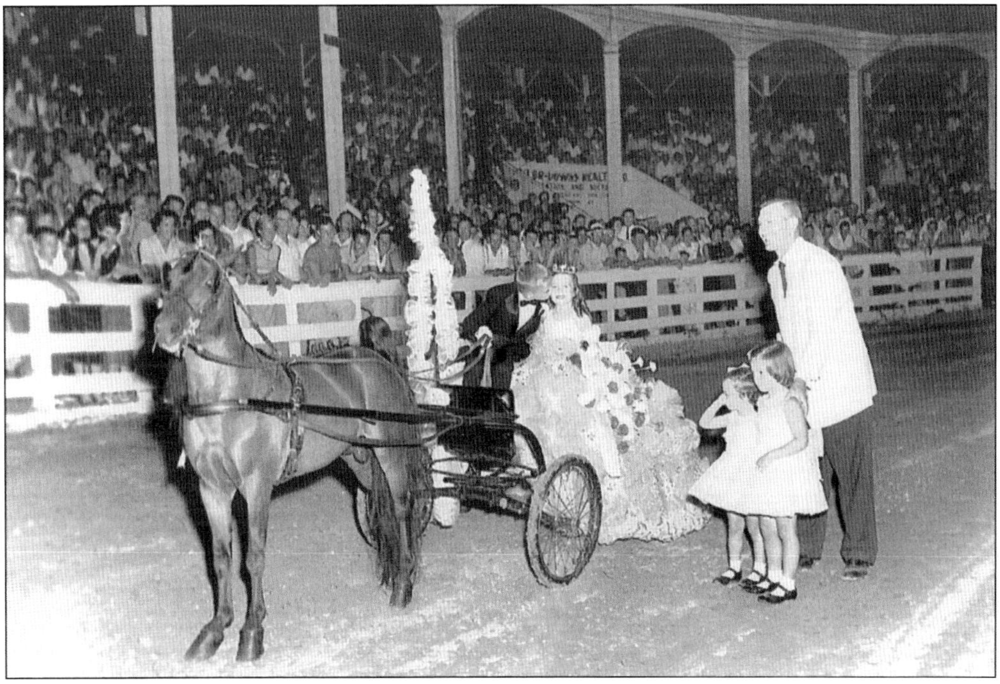

Jimmy Mountjoy served as ringmaster for many years at the Lawrenceburg Fair and Horse Show. Here, an enthusiastic crowd welcomes the start of the horse show. (Courtesy of Hanks collection.)

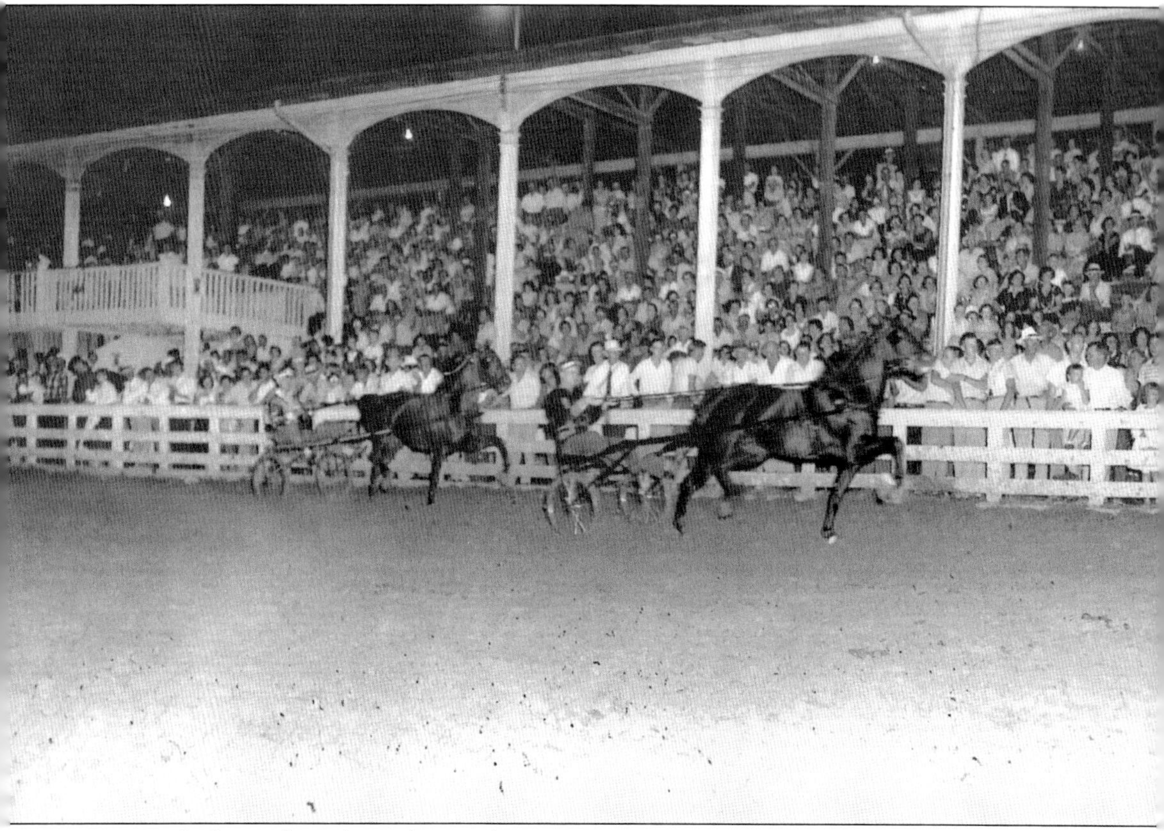

A packed grandstand watches as the trotters compete for first place at the Lawrenceburg Fair and Horse Show. The competition for prizes and prestige was great at the fair. (Courtesy of Hanks collection.)

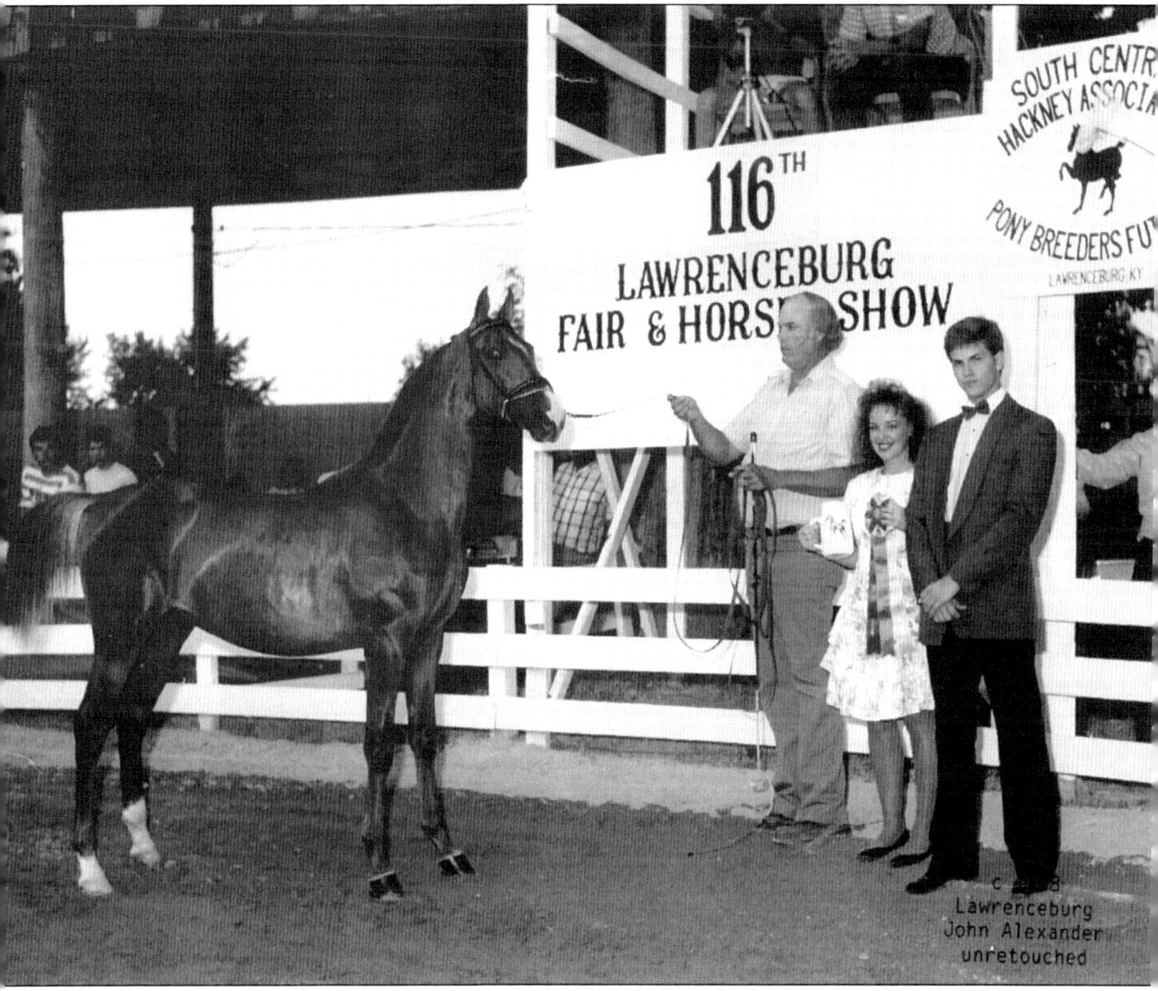

David Mountjoy (holding the reins), the grandson of W.D. and son of Billy Mountjoy, carries on the family tradition in the ring at the Lawrenceburg Fair and Horse Show. After the death of his father, David now operates the Mountjoy farm. (Courtesy of Hanks collection.)

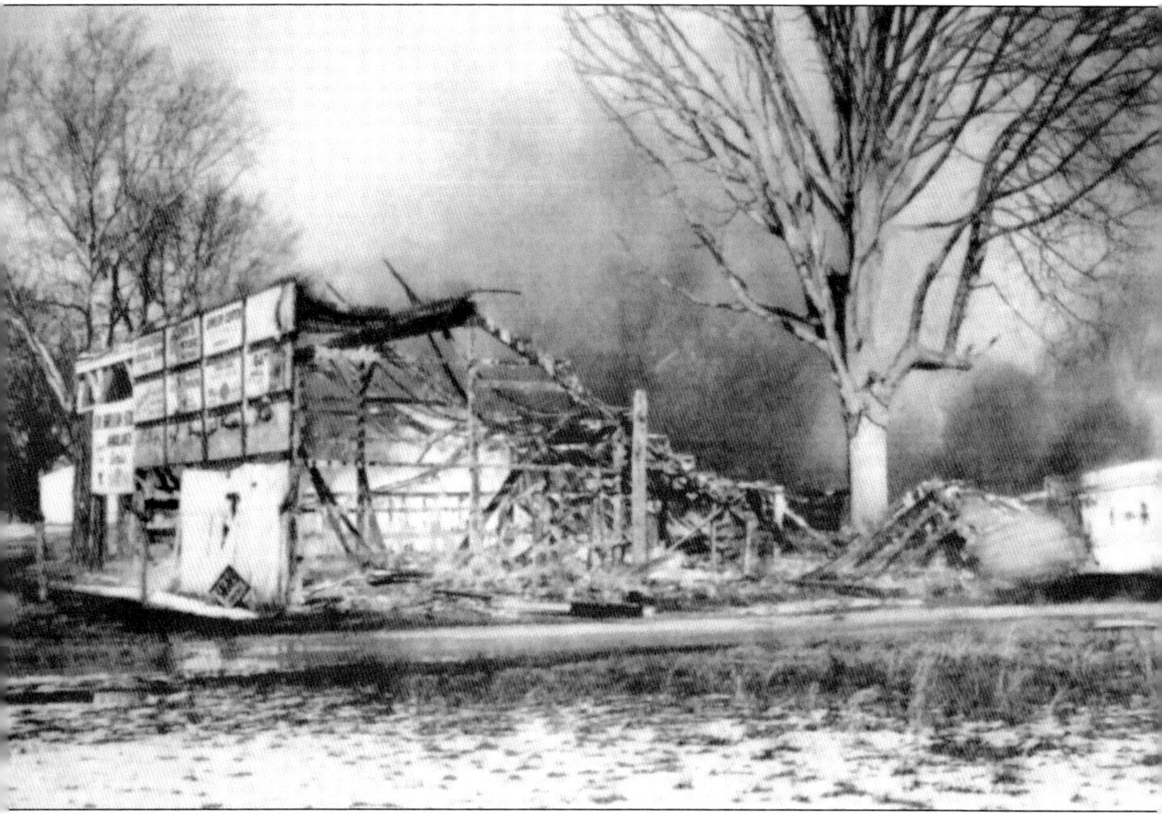

In 1961, fire destroyed the grandstand of the Lawrenceburg Fair and Horse Show. The structure was rebuilt, first without a roof, but one was added a few years later. This is the third grandstand for the Lawrenceburg Fair and Horse Show: the first was at McKee Woodland, the second is at the American Legion grounds, and third at the American Legion grounds also. (Courtesy of Hanks collection.)

Five

Other Industries

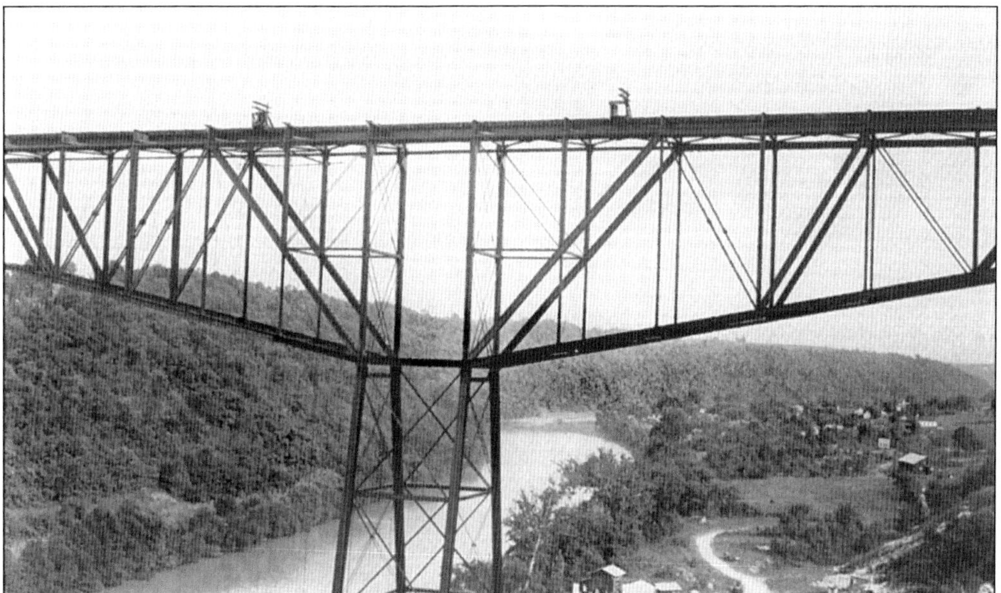

Young's High Bridge, which spans the Kentucky River, was named for Col. Bennett H. Young, who built the Southern Railroad through Anderson County in 1888. It is one of the great cantilever bridges in America. The span, along with the Blackburn Bridge, bypassed Tyrone and led to the decline of that village. Railroad building in the late 1880s started a fresh construction boom in Lawrenceburg, with many new buildings springing up. (Courtesy of McAnly collection.)

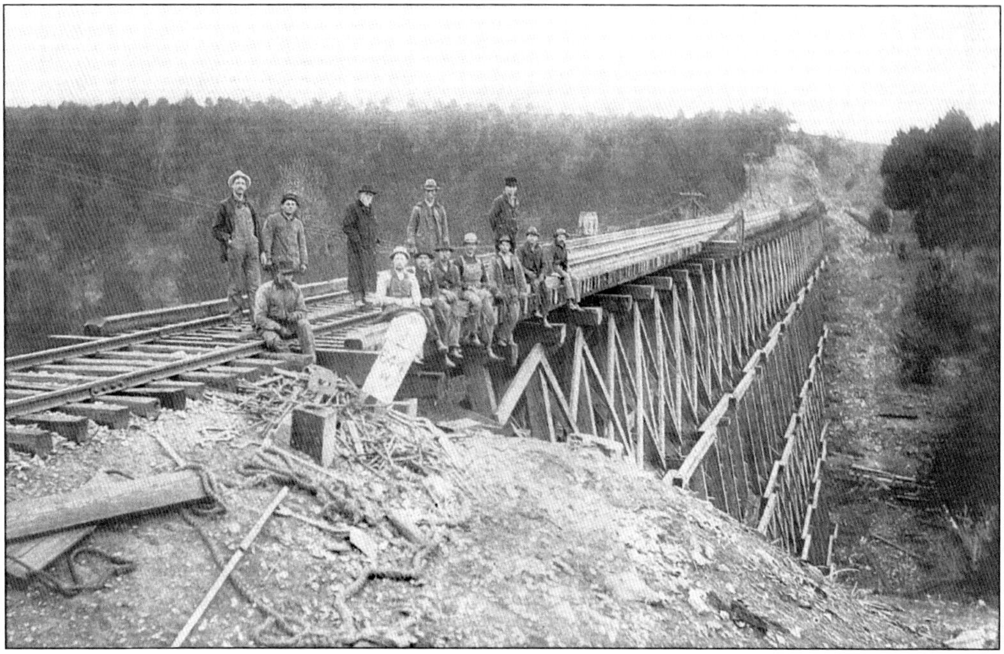

A railroad trestle spanned Cedar Brook Creek between Lawrenceburg and Tyrone. This railroad work crew is just finishing work on the new bridge. (Courtesy of Maxine Hawkins.)

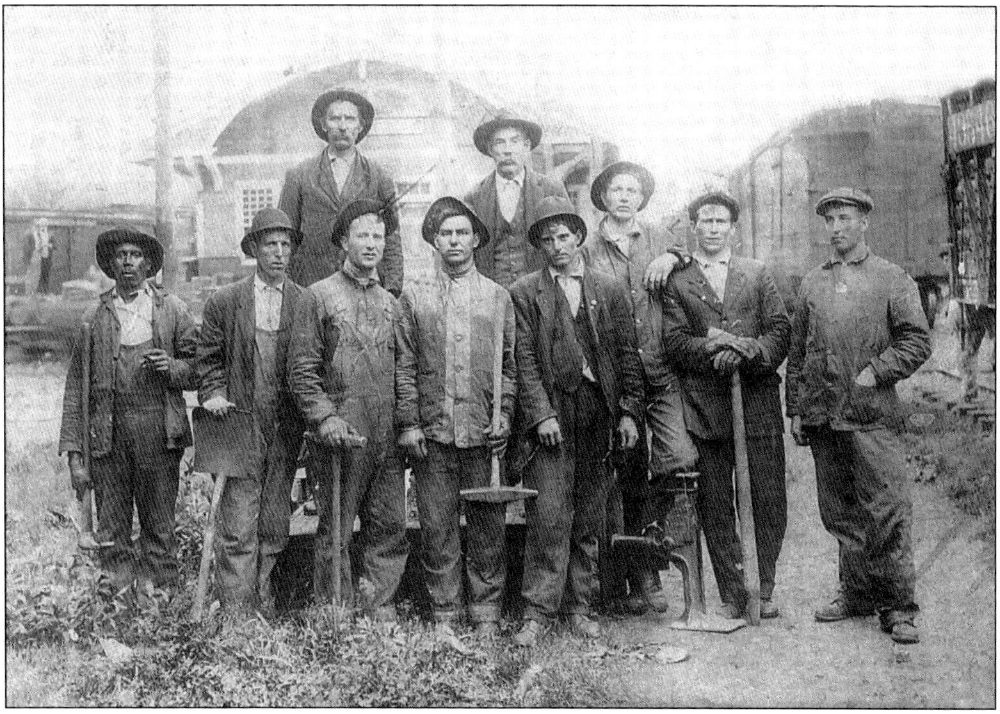

Railroad work crews laid track, built bridges, and cleared rights-of-way. The Southern Railroad opened up trade with northern cities, especially Cincinnati and towns in Kentucky and Tennessee. The competition for southern markets was especially intense between Cincinnati and Louisville. (Courtesy of Maxine Hawkins.)

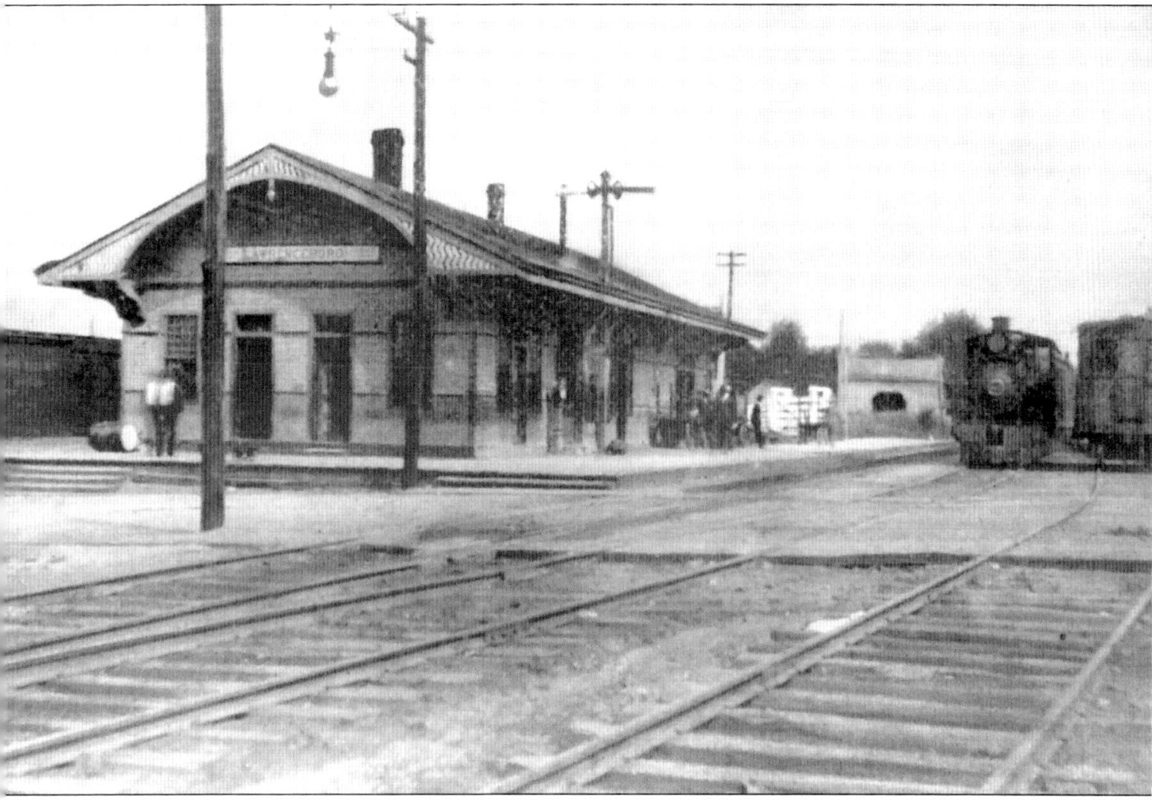

The Lawrenceburg depot was located off Court Street near the center of town. Incoming and outgoing mail and freight started or ended here. Passenger trains also arrived and departed from the depot. It was later moved to the edge of town, where it now serves as a produce market. (Courtesy of McAnly collection.)

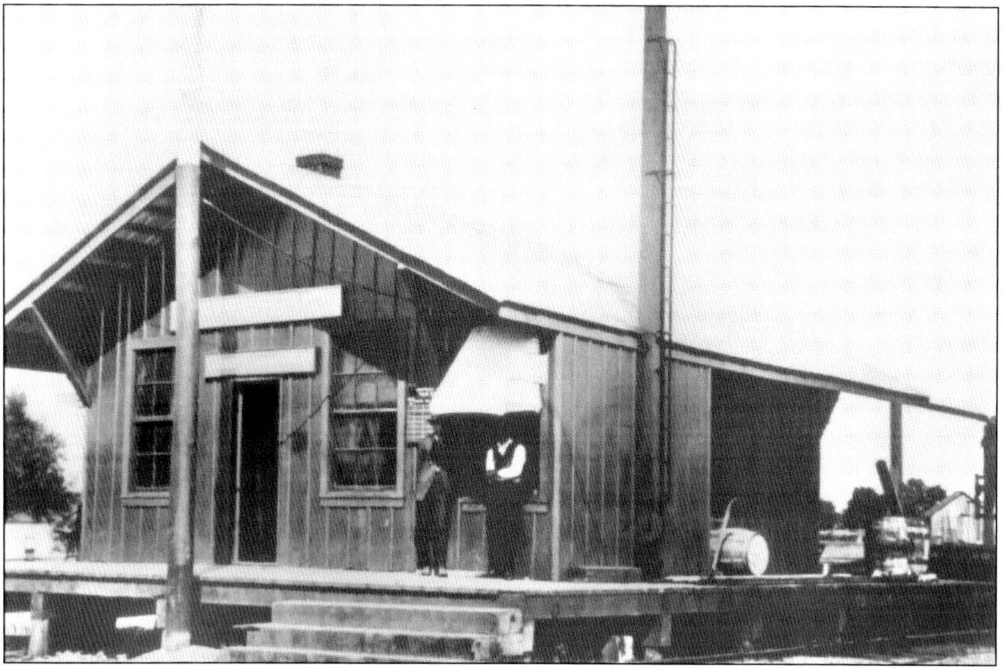

Railroads not only connected distant cities but also the small towns in between. The Alton depot was a connection to Lawrenceburg before interstate highways and roads reduced the time and distance of travel. The small local depots were phased out in the early 1900s. (Courtesy of Moffett collection.)

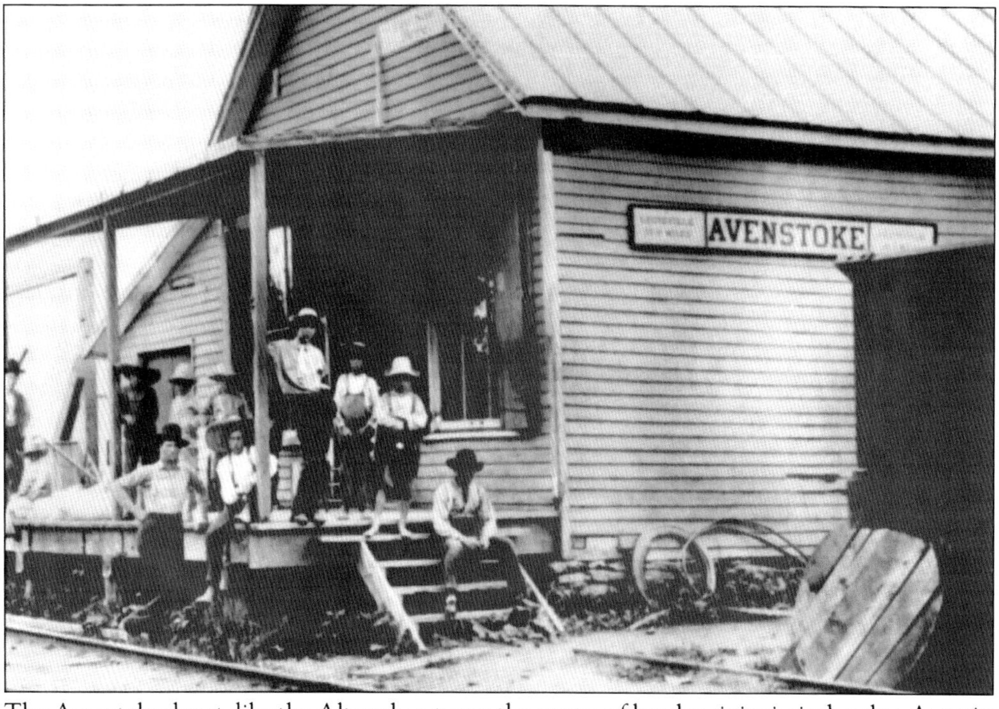

The Avenstoke depot, like the Alton depot, was the center of local activity in its heyday. As train traffic declined, small depots like Avenstoke disappeared. (Courtesy of Moffett collection.)

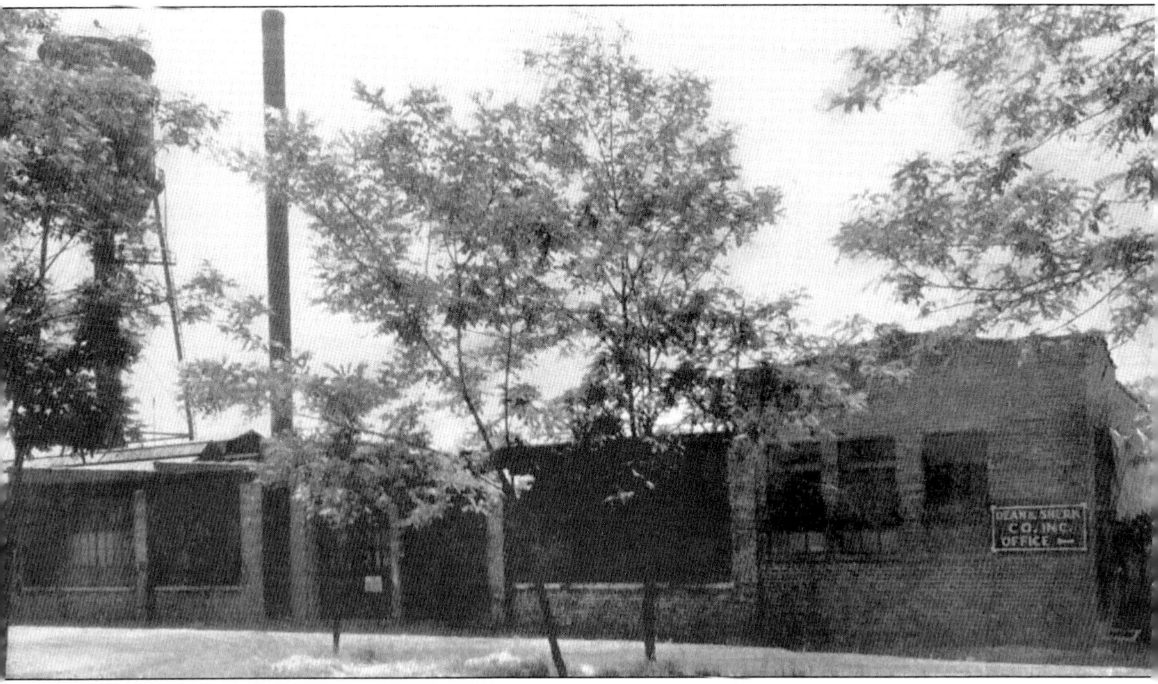

The Dean and Sherk Thread Factory was relocated from Detroit to Lawrenceburg in 1926 and began operation in 1927. The thread from this plant was distributed throughout the United States and to some foreign countries. Men and women worked in roughly equal numbers at the plant, with approximately 200 employees. (Courtesy of Anderson County Historical Society.)

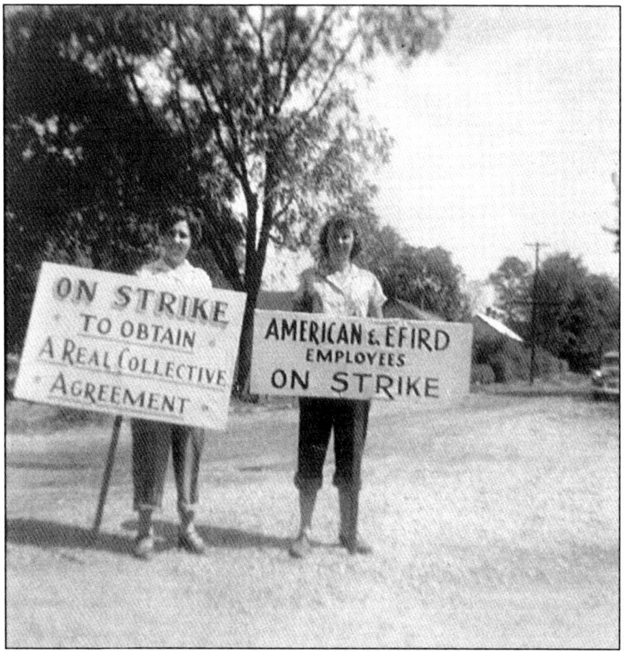

On June 26, 1952, American and Efrid Thread acquired Dean and Sherk, but labor problems broke out shortly thereafter. The International Ladies Garment and Workers Union attempted to organize the factory and managed to win a narrow election on October 15, 1952. (Courtesy of Jean Hawkins.)

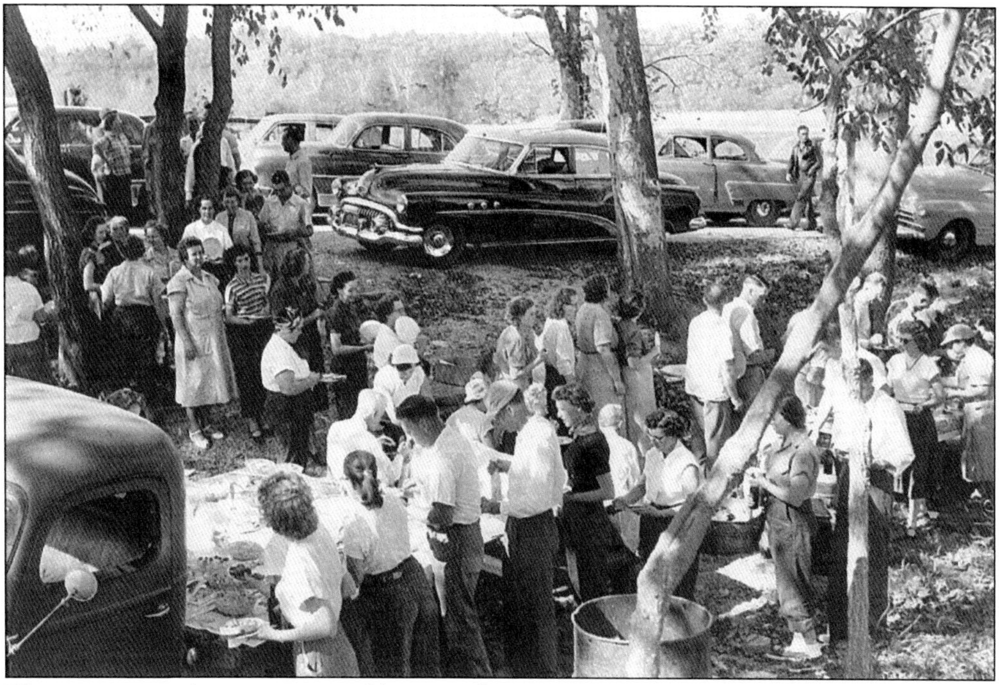

The company and union were unable to reach an agreement on the contract, and on July 20, 1953, employees staged a wildcat strike (without authorization of the union). Problems escalated between pro-union and pro-company employees. Driveways were covered with roofing nails, windows were broken at the factory and at nearby homes, cars were stopped, and people were threatened. American and Efrid Thread decided to close the operation and move to Mount Holly, North Carolina. The Lawrenceburg building was closed on December 18, 1953. (Courtesy of Jean Hawkins and Geraldine Trent.)

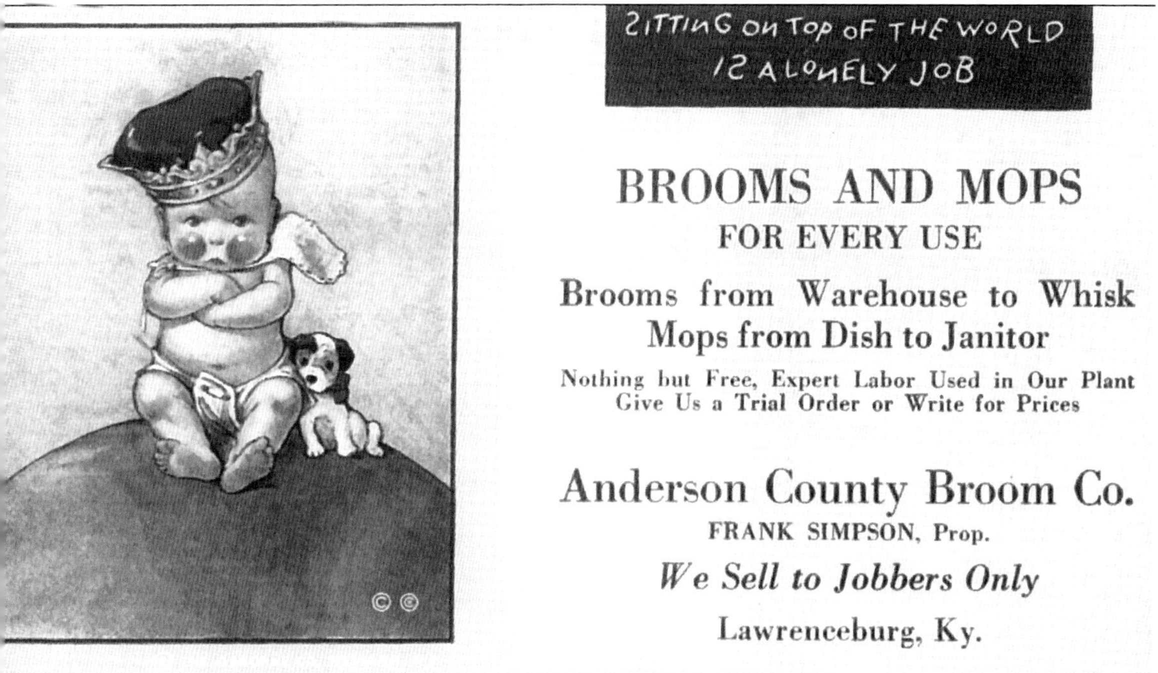

Two broom factories called Lawrenceburg home from the 1920s through the 1960s. The Anderson County Broom Company was first opened by Jack Parlin of Cincinnati near Tyrone and was known as the Riverside Broom Works. In 1920, Parlin sold the business to Frank Simpson and Carl Cooper, who moved the business to Lawrenceburg. Later, the factory was bought by Fannie Gaines. (Courtesy of Hawkins collection.)

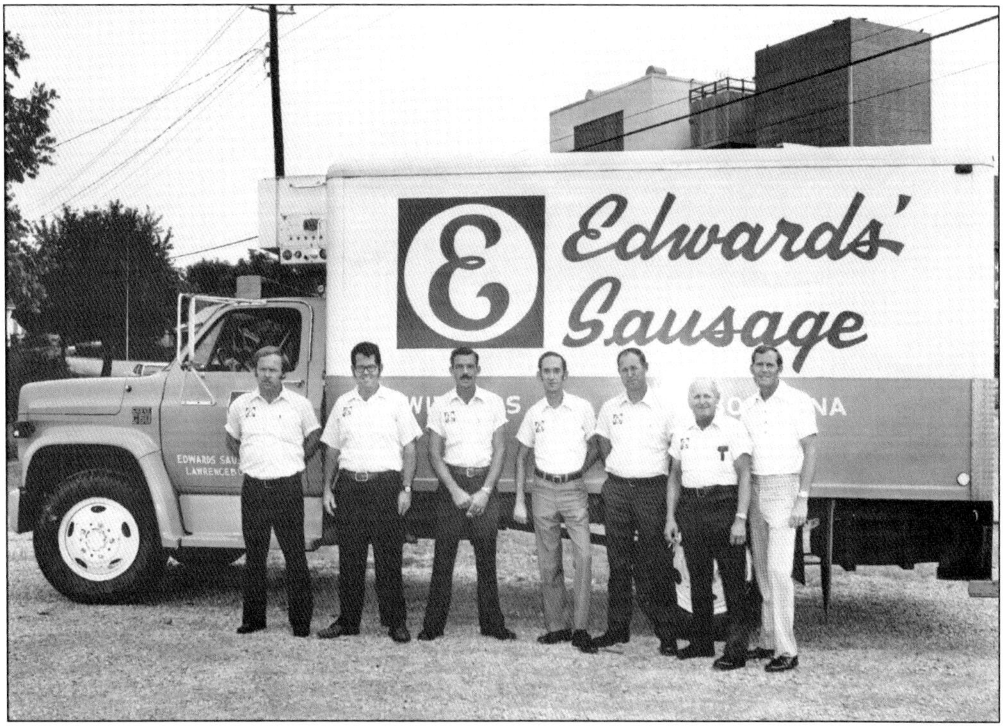

Edwards' Sausage of Lawrenceburg began in Harrodsburg in 1936. In 1939, C.M. Edwards and his son Marvin formed a partnership and moved to Lawrenceburg. They built a small plant and, in the early days, delivered sausage to four counties: Anderson, Mercer, Jessamine, and Franklin. In 1941, delivery routes were expanded and the company grew. An Edwards' delivery truck is seen here with drivers who have received safe-driving awards. They are, from left to right, Paul Ewalt, Stave Trent, Don Cooper, Dudley Dennis, Sidney Puckett, Edwin Springate, and Jim Link. (Courtesy of Link collection.)

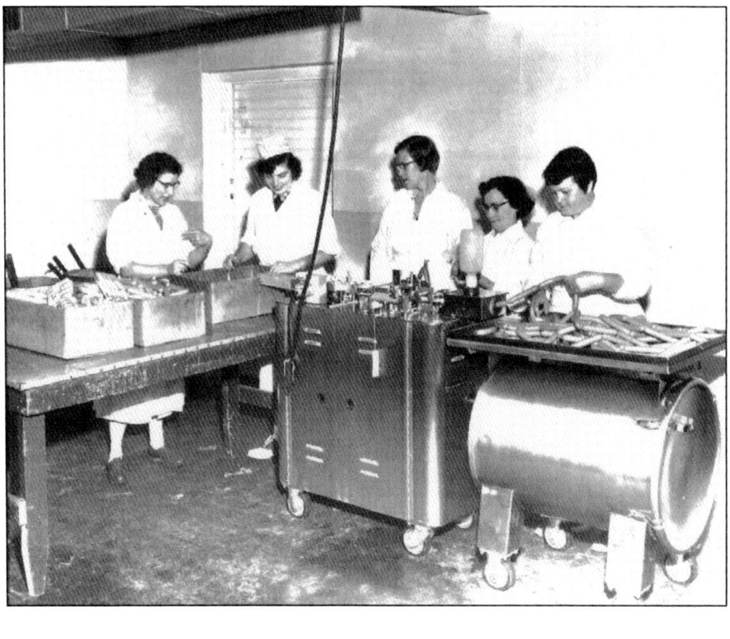

Edwards' made sausage and other associated products. Seen here on the hot-dog assembly line are, from left to right, Gladys Barnett, Della Blackwell, Loretta Wells, Beatrice Fritz, and Mildred Goodlett. (Courtesy of Hawkins collection.)

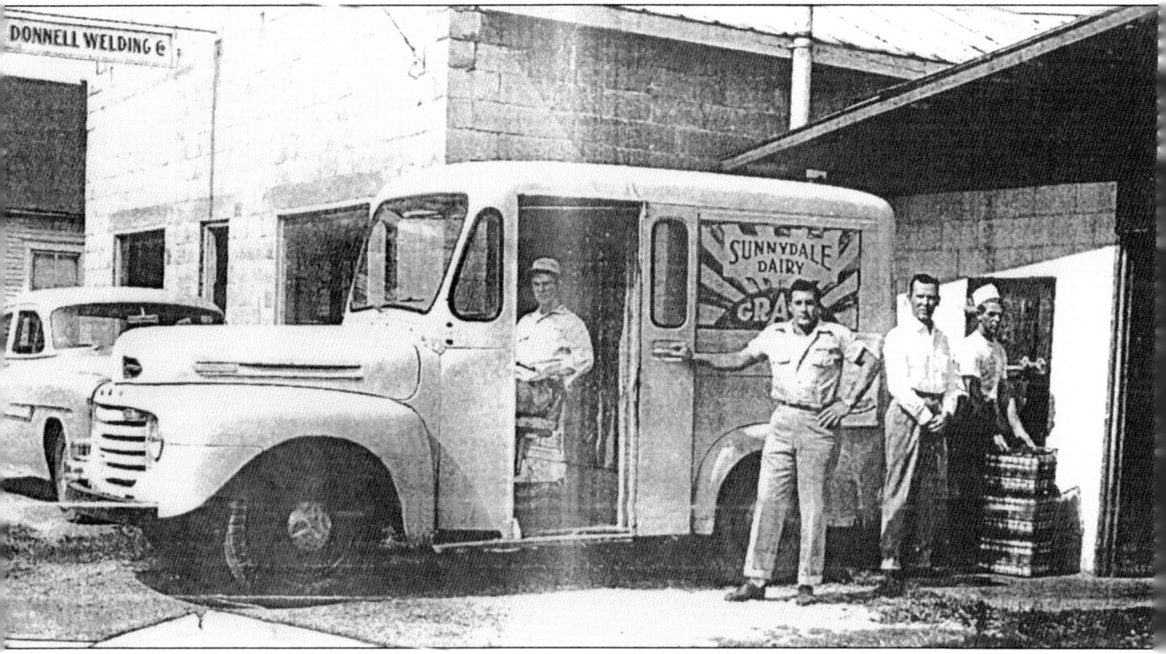

In 1955, the Sunnydale Dairy was going strong. The processing plant was on Court Street, and daily deliveries were made by the staff. This delivery truck is pictured being loaded, and posing with it are driver Rollin Cole, co-owner Terry Campbell (second from right), and two unidentified men. (Courtesy of Rollin Cole.)

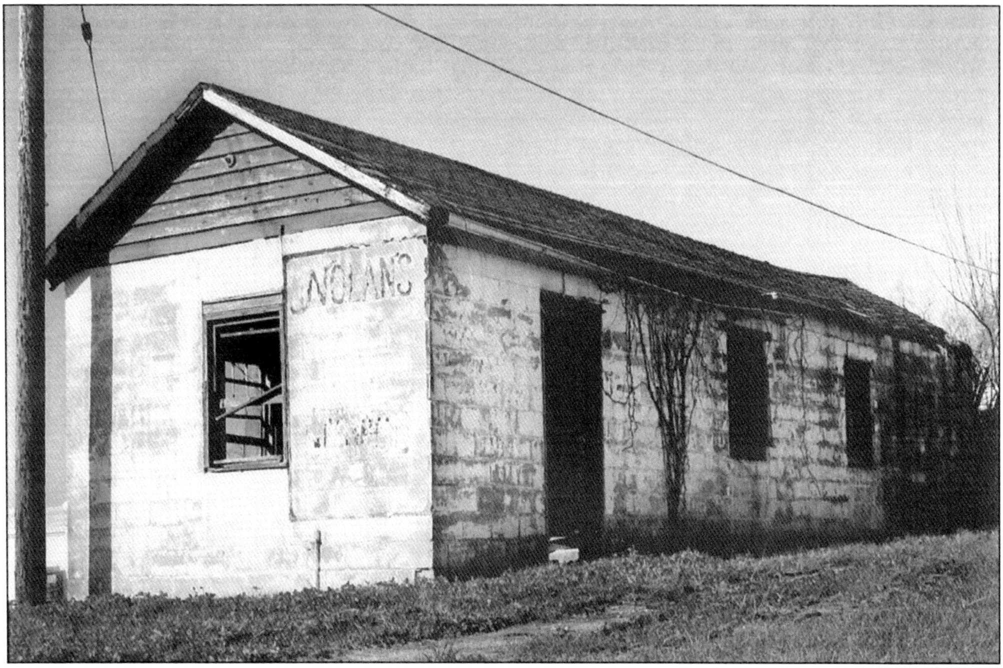

Lawrenceburg Transfer Company was begun by Mel Peach, who later sold the business to Tom B. Ripy and his partners. The old office of the Lawrenceburg Transfer Company gave way to a larger facility as the business prospered. The company had offices in Danville, Louisville, and Lawrenceburg and delivered all over the state. The former office on Court Street is seen here. (Courtesy of Sayre collection.)

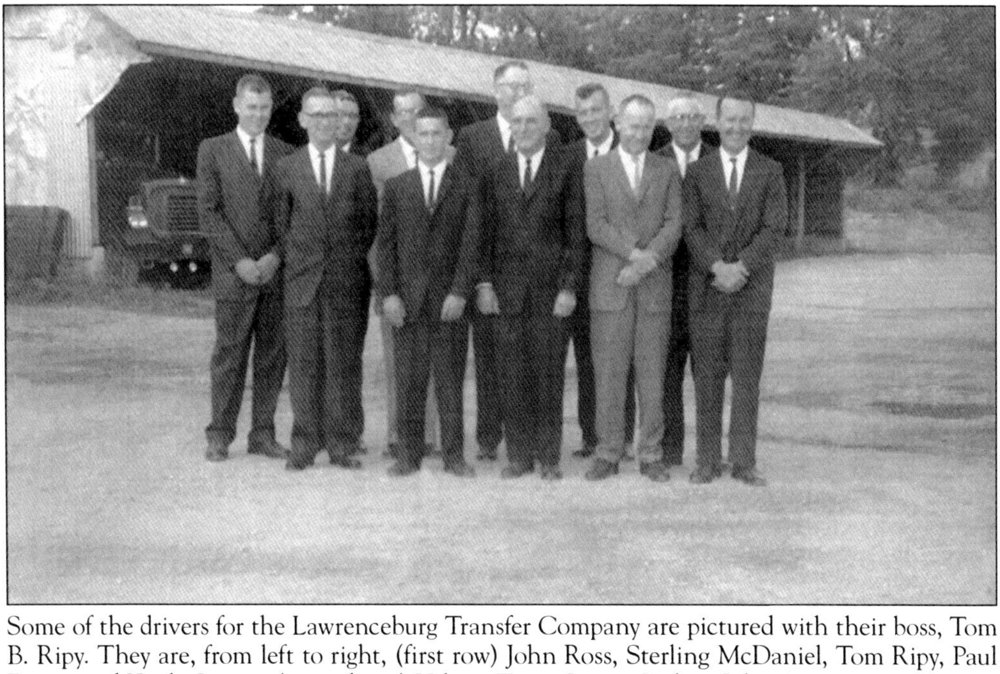

Some of the drivers for the Lawrenceburg Transfer Company are pictured with their boss, Tom B. Ripy. They are, from left to right, (first row) John Ross, Sterling McDaniel, Tom Ripy, Paul Drury, and Keith Gaines; (second row) Hiliary Trent, James Catlett, John Sayre, J.O. Newman, Jim Sayre, and Chester McGaughey. (Courtesy of Sayre collection.)

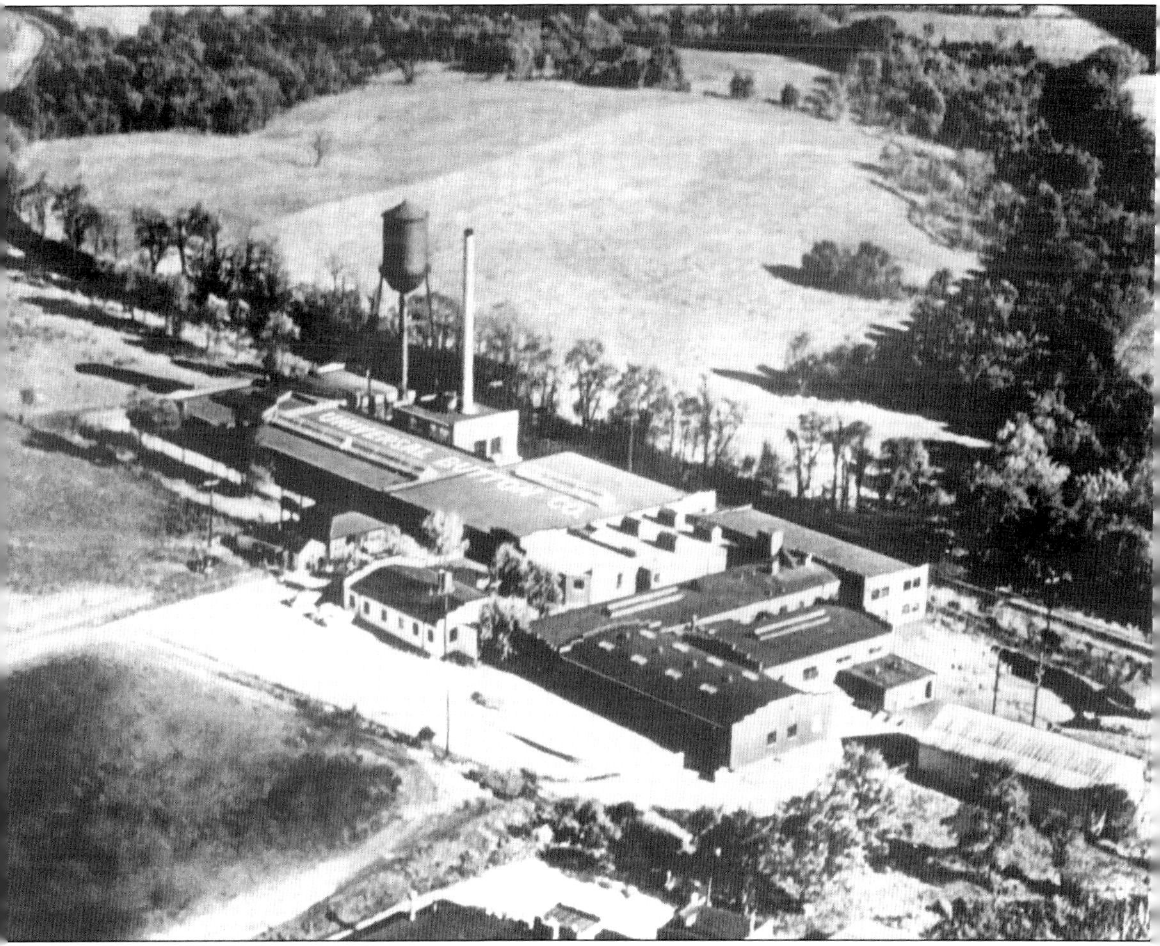

On June 23, 1955, Talon, Inc. (Universal Button Company) agreed to lease the former thread factory for 10 years providing certain repairs were made. The Anderson County Industrial Corporation was chartered on June 29, 1955, to serve as a nonprofit entity to facilitate the purchase of the property from American and Efrid Thread Company and act as the leasing authority. American and Efrid sold the building and acreage to the Anderson County Industrial Corporation for $162,000, after which the Universal Button Company relocated to Lawrenceburg. (Courtesy of McEntyre collection.)

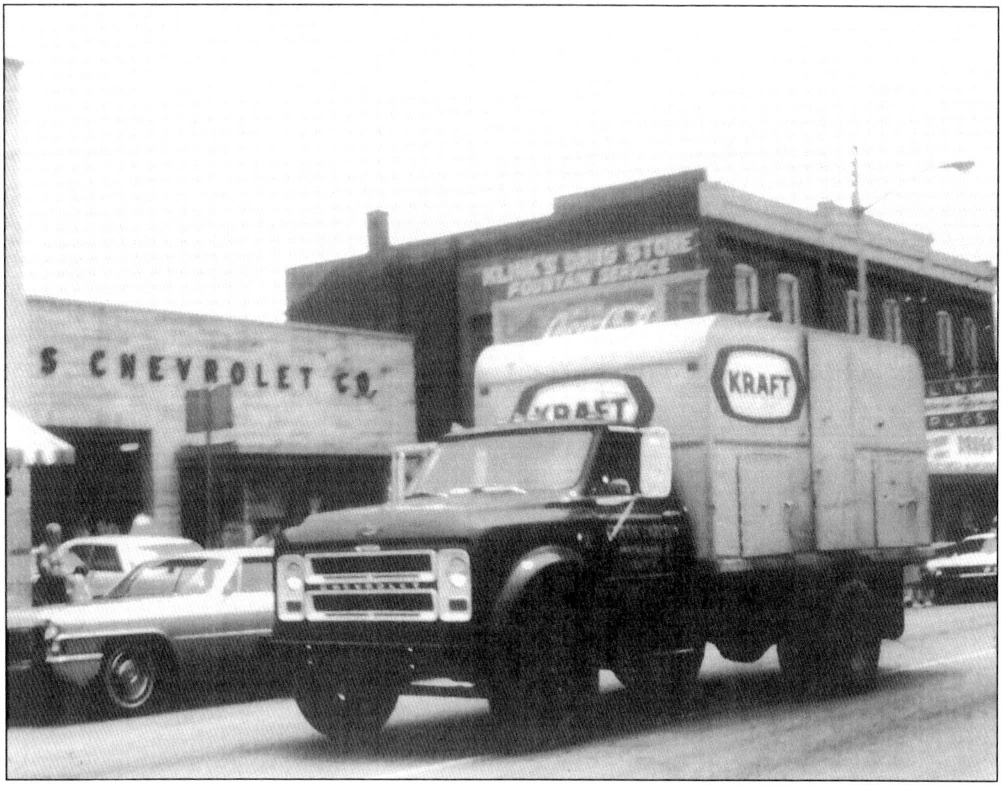

Revenue was needed by Anderson County after Prohibition closed the distilleries. In 1927, a group from Lawrenceburg went to Paoli, Indiana, to try to bring Kraft Cheese Company to the county. The company needed 2,000 cows and required the town to pledge to have that many available to provide milk. The plant was erected and opened in 1928 under the name of Kraft-Phenix Cheese Company. (Courtesy of Betty Hazelwood.)

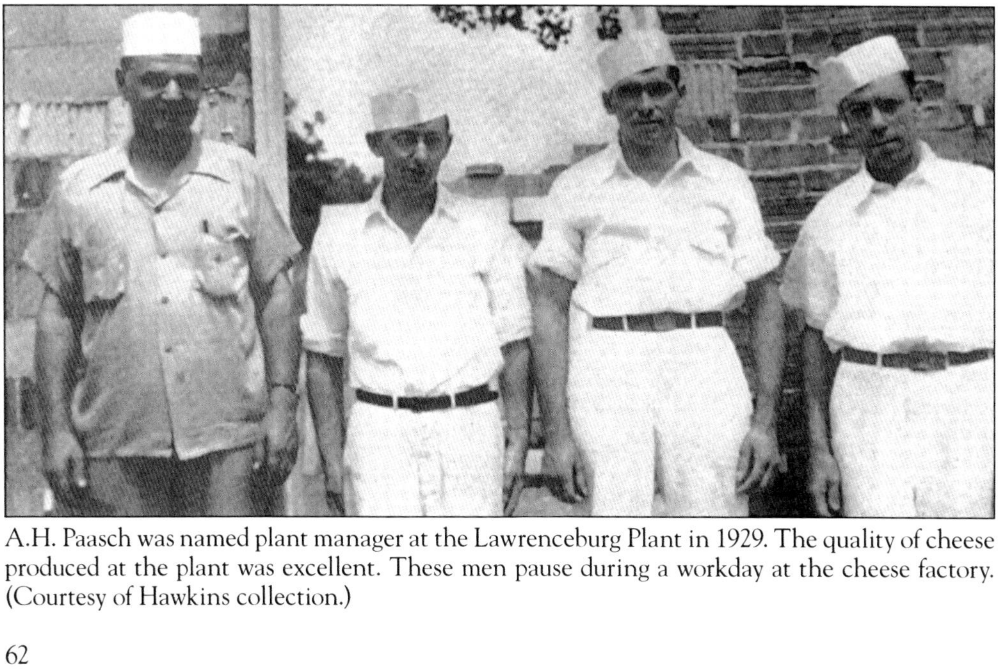

A.H. Paasch was named plant manager at the Lawrenceburg Plant in 1929. The quality of cheese produced at the plant was excellent. These men pause during a workday at the cheese factory. (Courtesy of Hawkins collection.)

Six

Day-to-Day Life

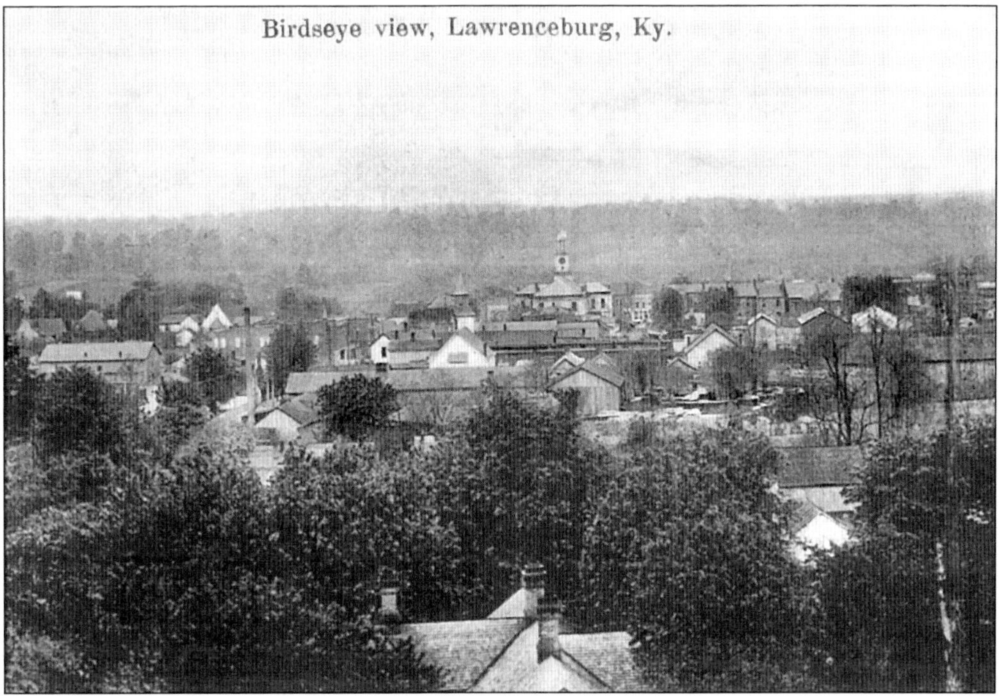

The small, quiet town of Lawrenceburg is seen here in a bird's-eye view of rooftops in 1915. Together, the people who lived in these houses formed a close-knit community through their interactions in church, business, leisure, and civic affairs. (Courtesy of McAnly collection.)

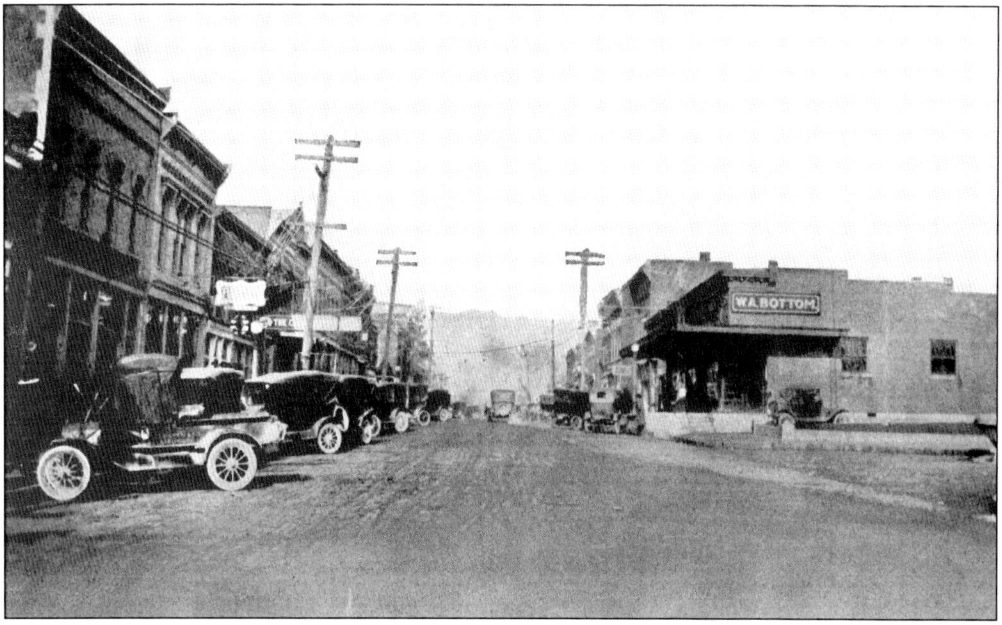

Automobiles are parked along Main Street in downtown Lawrenceburg, proving that it was a place to be. W.A. Bottom Saloon (on the right) is at a central location on Main Street in this early-1900s image. Bottom, a Mercer County native, came to Lawrenceburg in his early 20s. He maintained business interests in Tyrone, Lawrenceburg, and throughout the state, becoming one of Lawrenceburg's leading citizens and businessmen. (Courtesy of McAnly collection.)

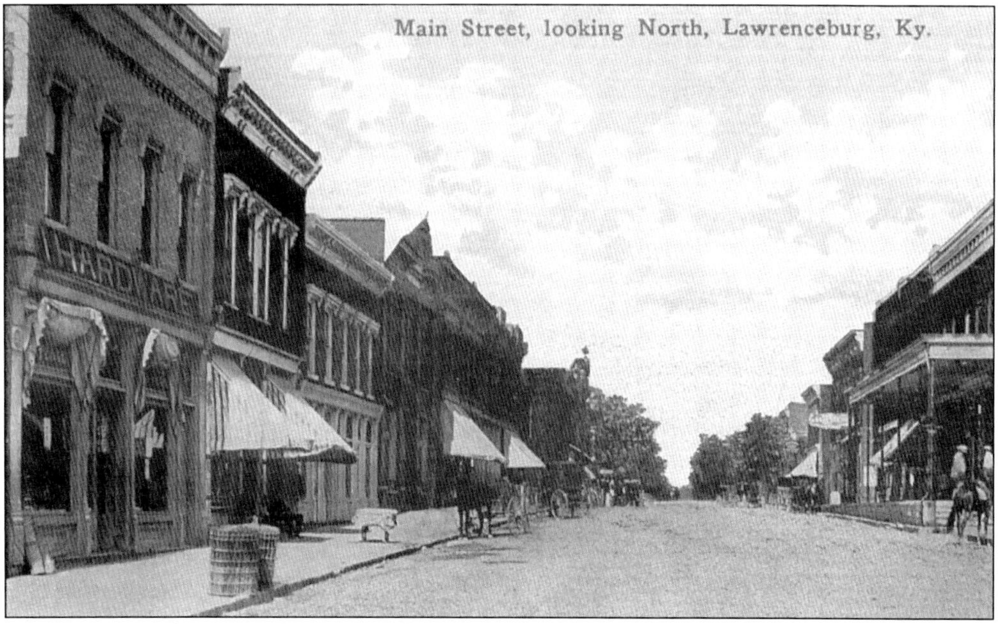

Another view of Lawrenceburg around 1900 shows prosperous storefronts and clean streets. In *History of Anderson County*, Lydia K. Bond described Lawrenceburg thusly: "The town itself presents an inviting appearance to the stranger. Main Street is clean, neat and presentable as any principal street of any town in the state. From whatever the direction Lawrenceburg is approached, no slums or unkempt buildings are encountered." (Courtesy of Hawkins collection.)

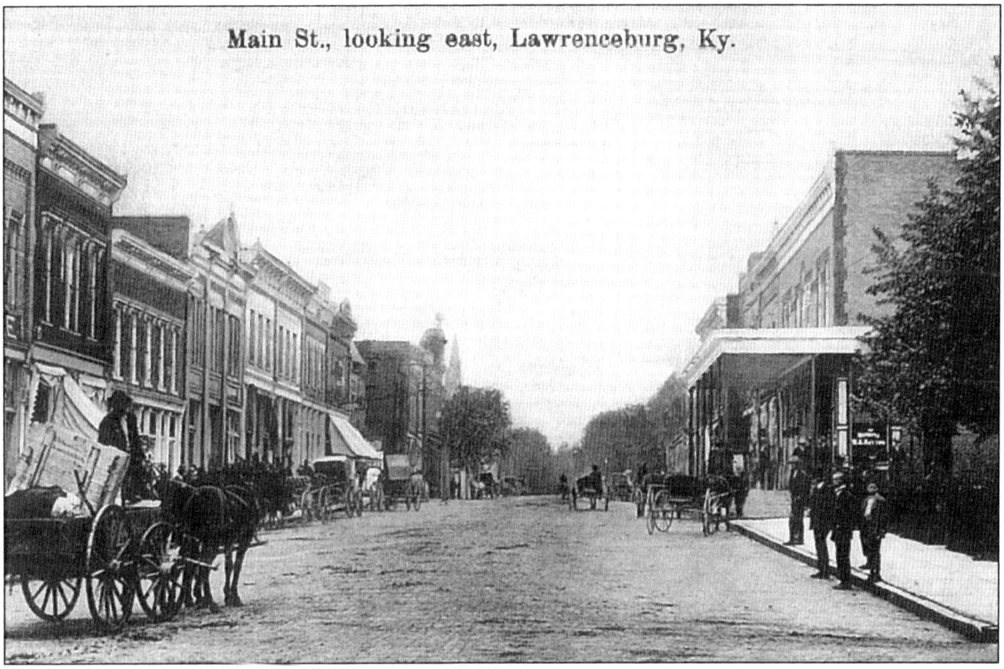

Seen here in the early 1900s, horse-drawn vehicles and automobiles vie for places to park. The tall steeple of the Lawrenceburg Baptist Church is on the left. (Courtesy of Hawkins collection.)

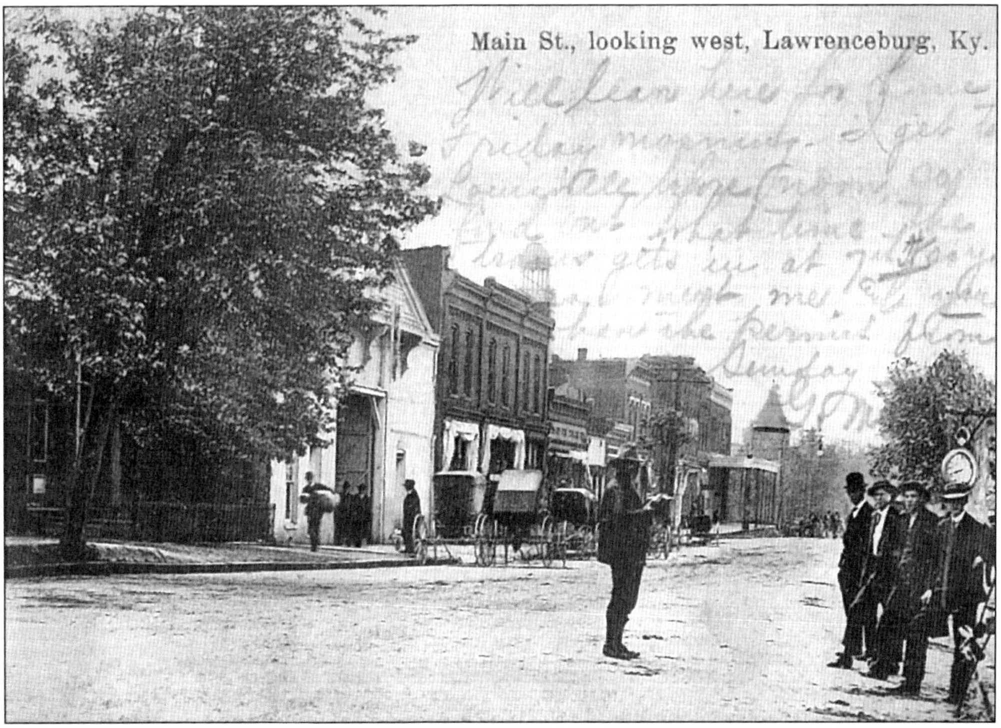

This 1915 view of Main Street looks west, showing the courthouse in the background and the First Christian Church spire at its former location at Main and Court Streets. (Courtesy of Hawkins collection.)

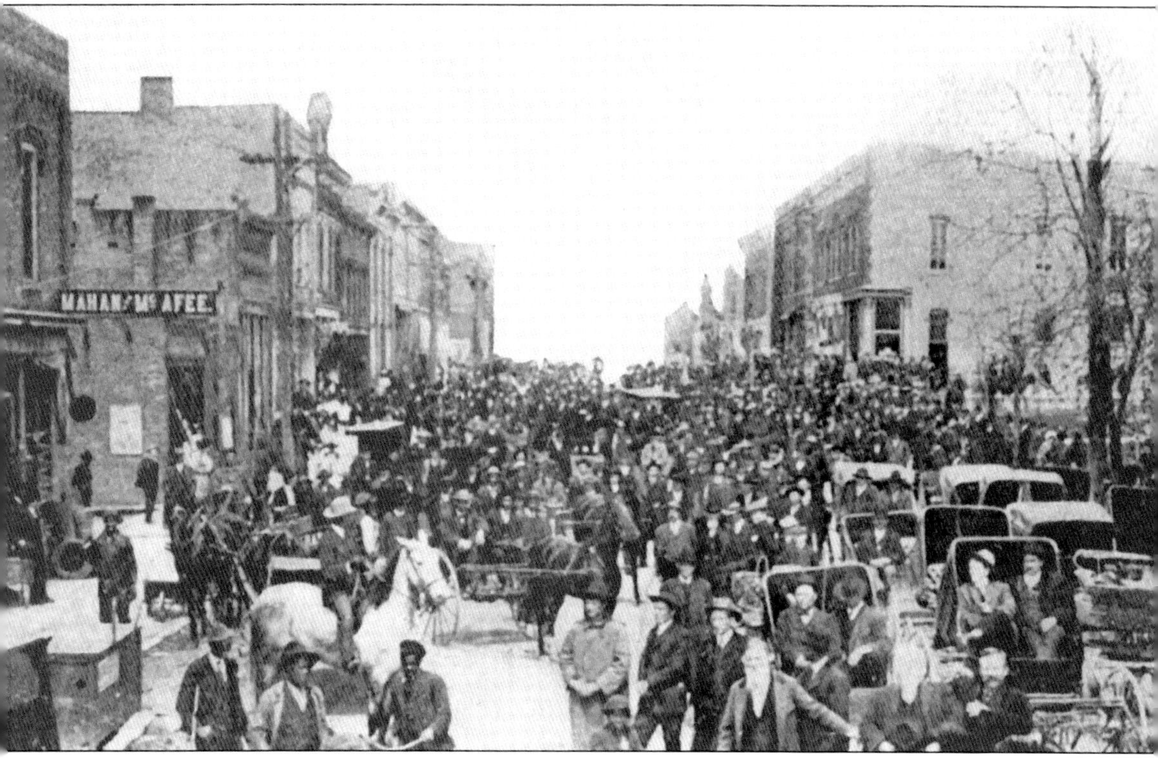

Court Day was also a social day in downtown Lawrenceburg. In this 1909 image, Main Street is filled with people and automobiles. (Courtesy of Hawkins collection.)

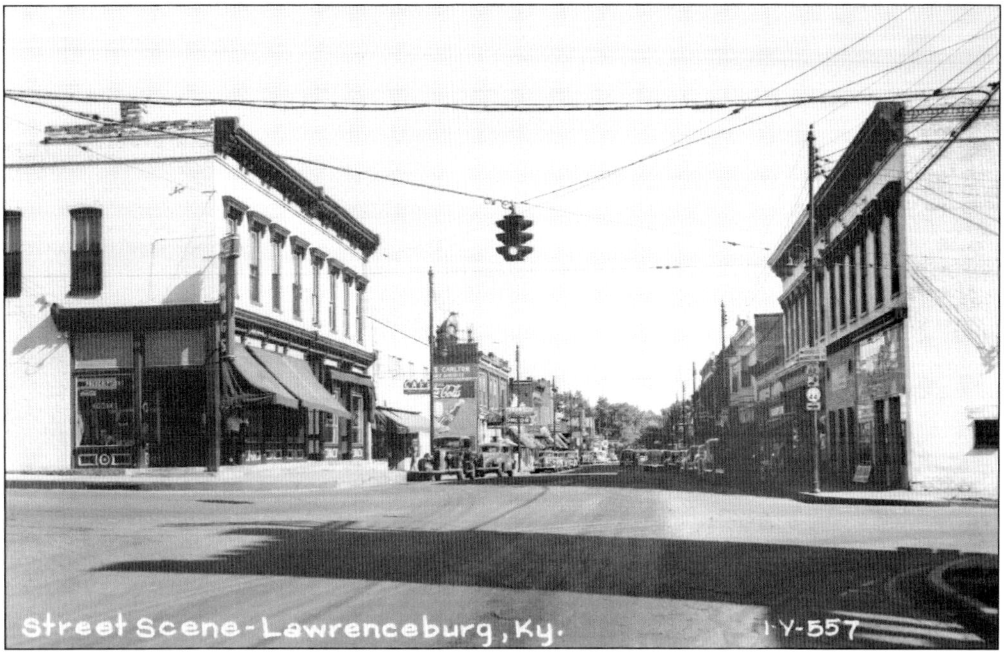

By the 1940s, changes were evident. Ballard's Drugs still stood on the corner of Main Street, but there was a stoplight hanging over the middle of the road. The Lyric Theatre is on the right. (Courtesy of McAnly collection.)

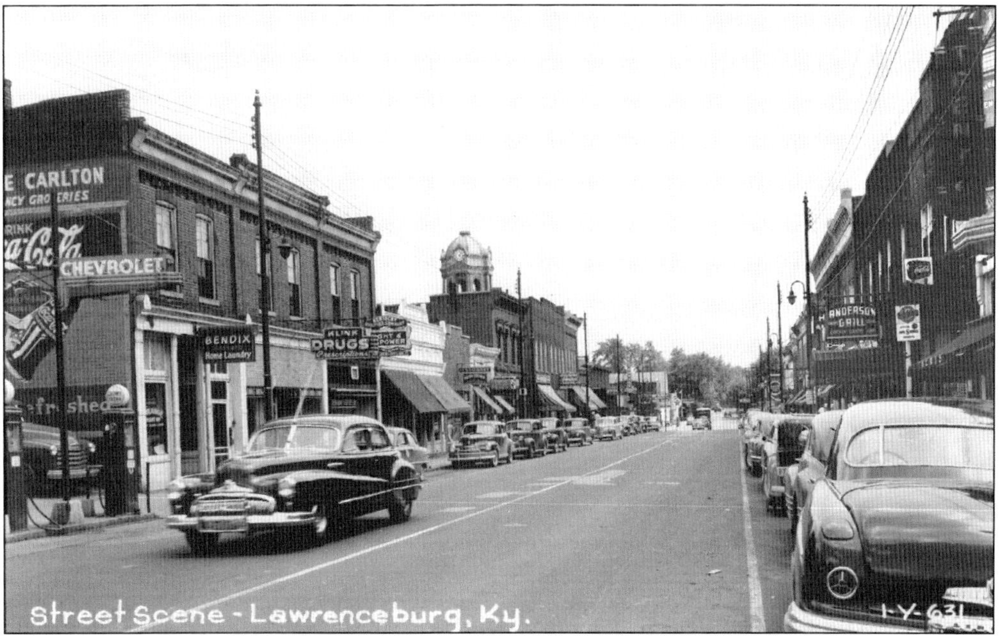

Also in the 1940s, Main Street included a Chevrolet dealership, Klink's Drugs, Kentucky Utilities, the Anderson Grill, and Trent's Clothing, among others. (Courtesy of Charlie Klink.)

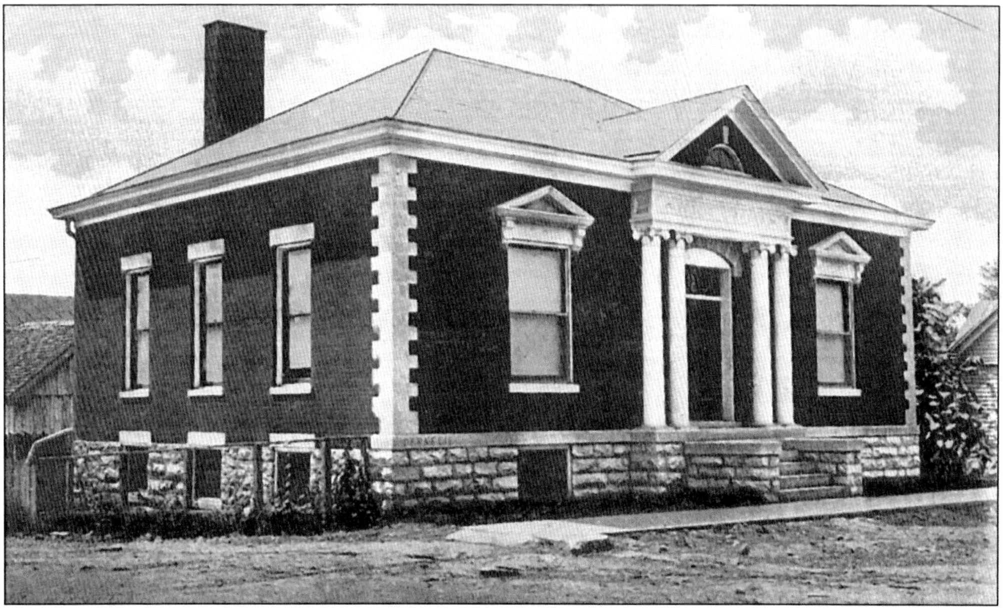

Andrew Carnegie funded the building of the Carnegie Library in Lawrenceburg. In addition, he also provided funds for some of the furnishings. This building now houses the Anderson County Historical Museum. (Courtesy of Hawkins collection.)

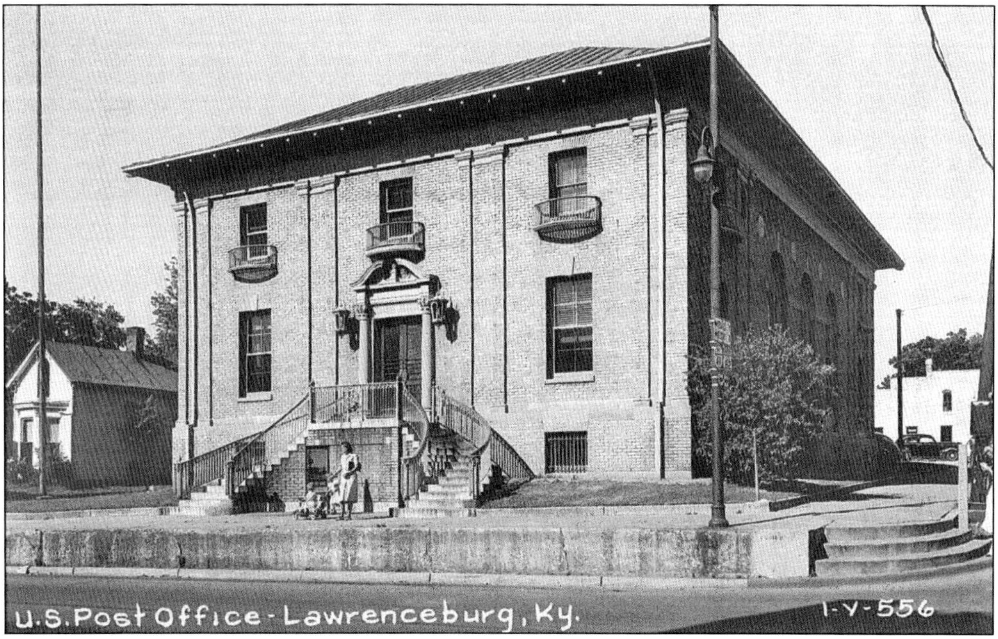

The Lawrenceburg Post Office was built on the site of the former Galt house. The post office housed the Selective Service System and the Soil Conservation Service for a number of years. Today, the former post office is the city building, and the post office has relocated to West Broadway. (Courtesy of McAnly collection.)

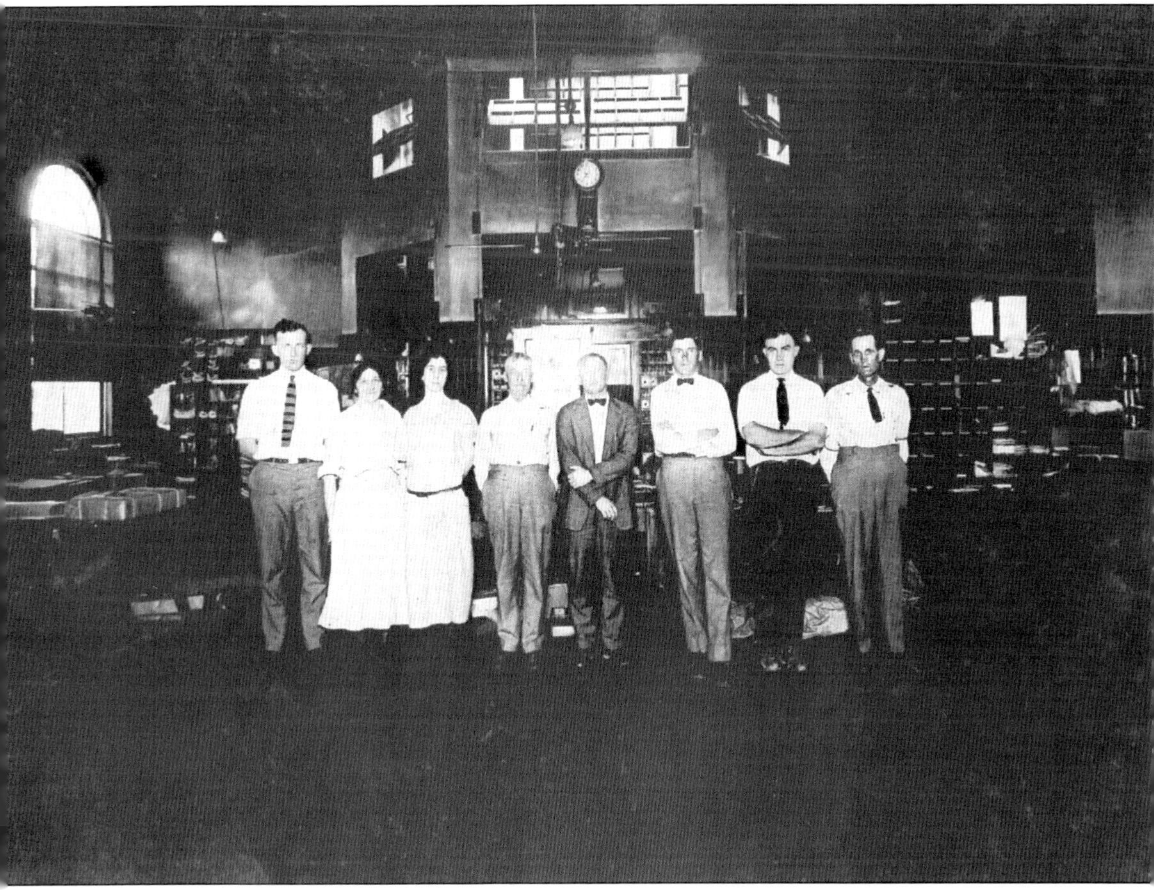

The interior of the post office is seen here around 1930. The postal workers posing are, from left to right, Bill Nevins, Kitty Searcy, Vena Bryant, unidentified, Bert Searcy, Tom Stratton, Mitch Bailey, and unidentified. (Courtesy of Bryant collection.)

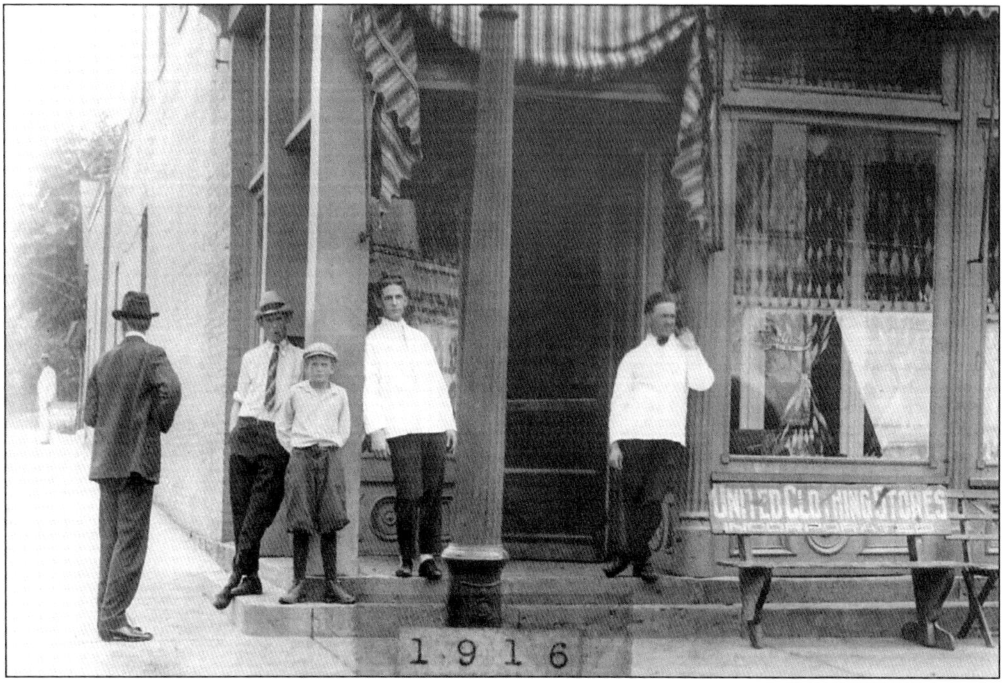

Ballard's Drug Store has long been located at the corner of Main and Woodford Streets, which is known as "Ballard's Corner." It is one of Lawrenceburg's oldest businesses, operating in the same location since the 1880s. (Courtesy of Geoghegan collection.)

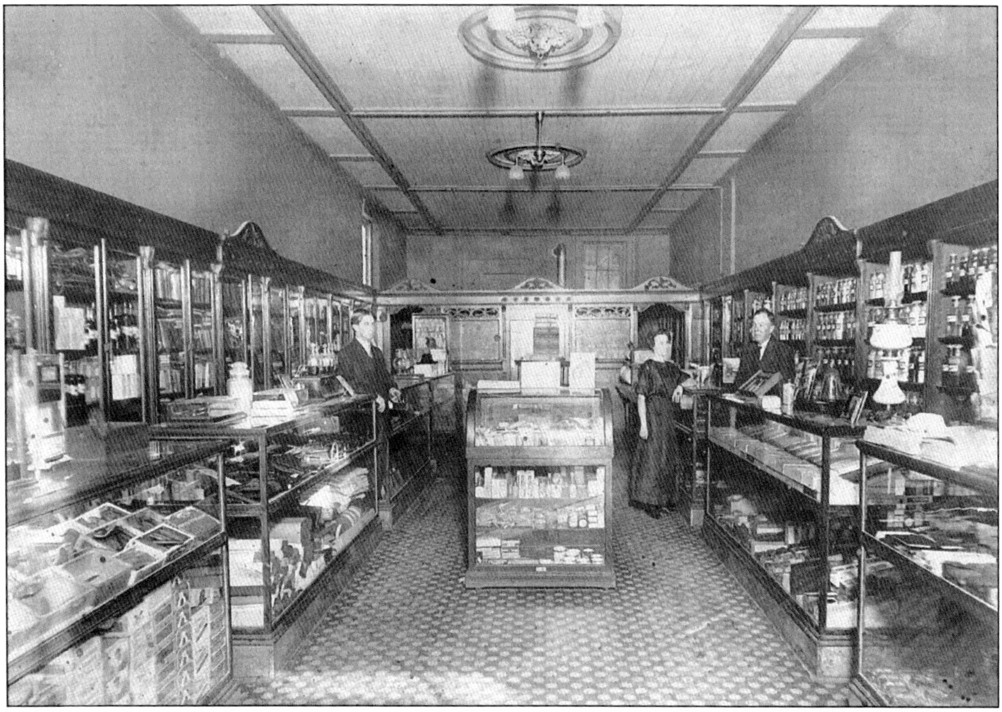

The interior of Ballard's Drug Store is seen here in 1902. At right are owners Gertrude and Thomas Ballard. (Courtesy of Geoghegan collection.)

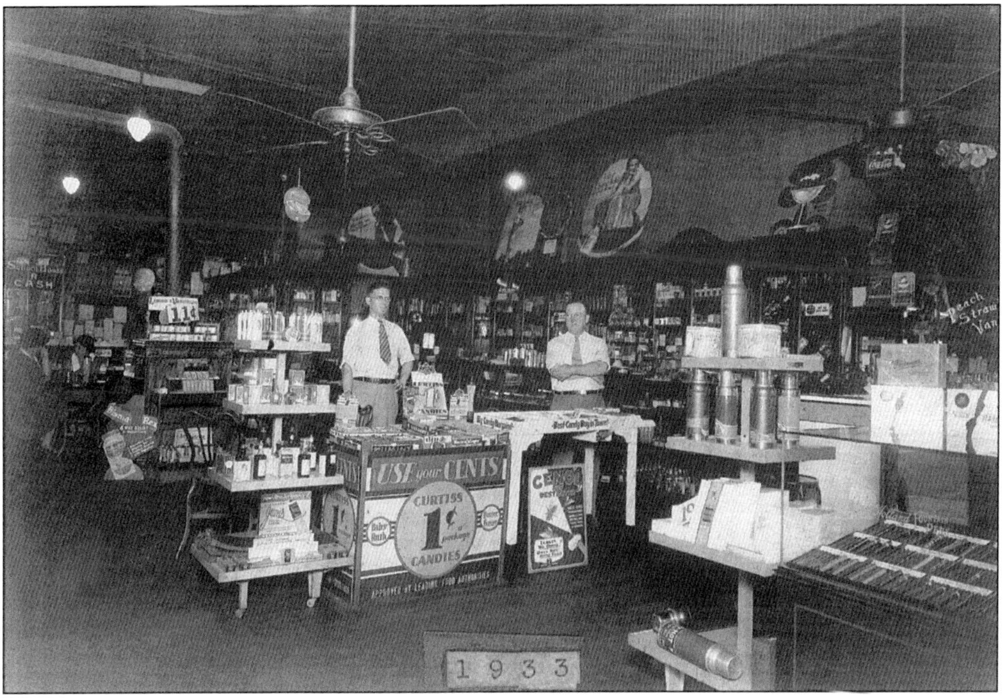

By 1933, the interior of Ballard's Drug Store had been through many changes and now offered many more products than just drugs. Seen here are Ben Young and Speed Woods, the latter of whom lived in an upstairs apartment. (Courtesy of Geoghegan collection.)

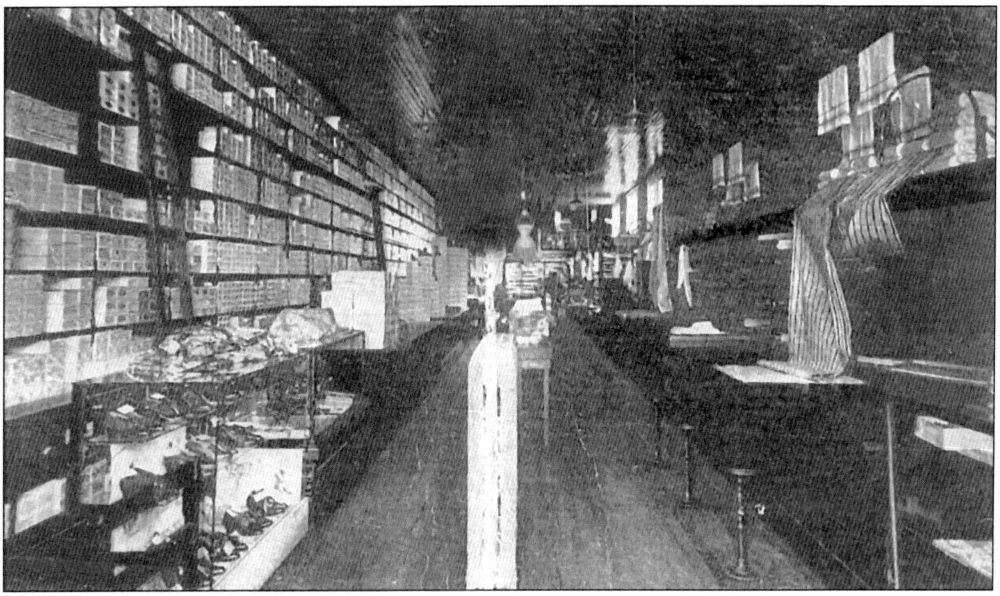
The S.B. Hawkins and Company store sold dry goods of all types. (Courtesy of Hawkins collection.)

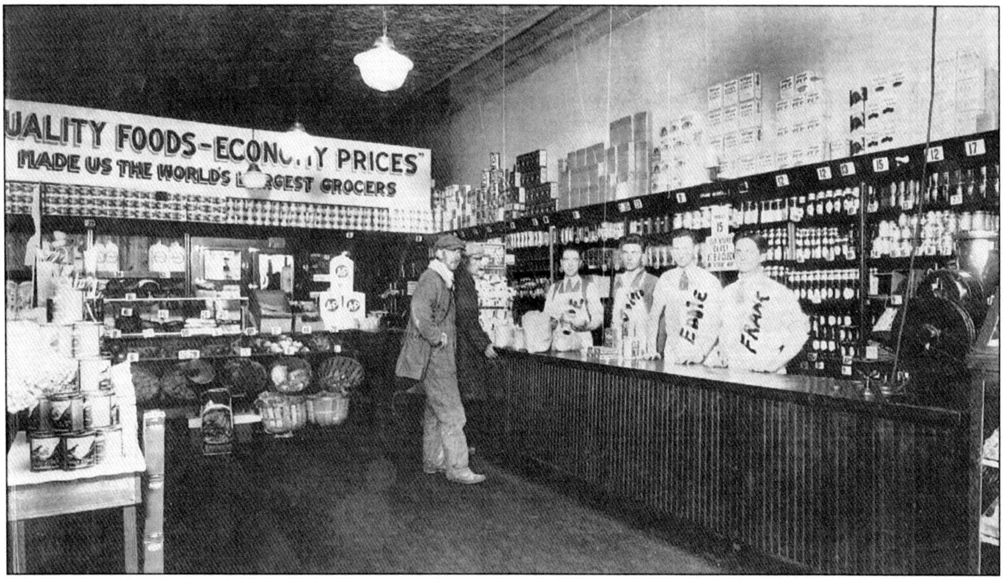

The A&P grocery store on Main Street is pictured in 1928. The two clerks who can be identified here are John Lewis Elliott (second from left) and Eddie Morris (second from right). (Courtesy of Morris collection.)

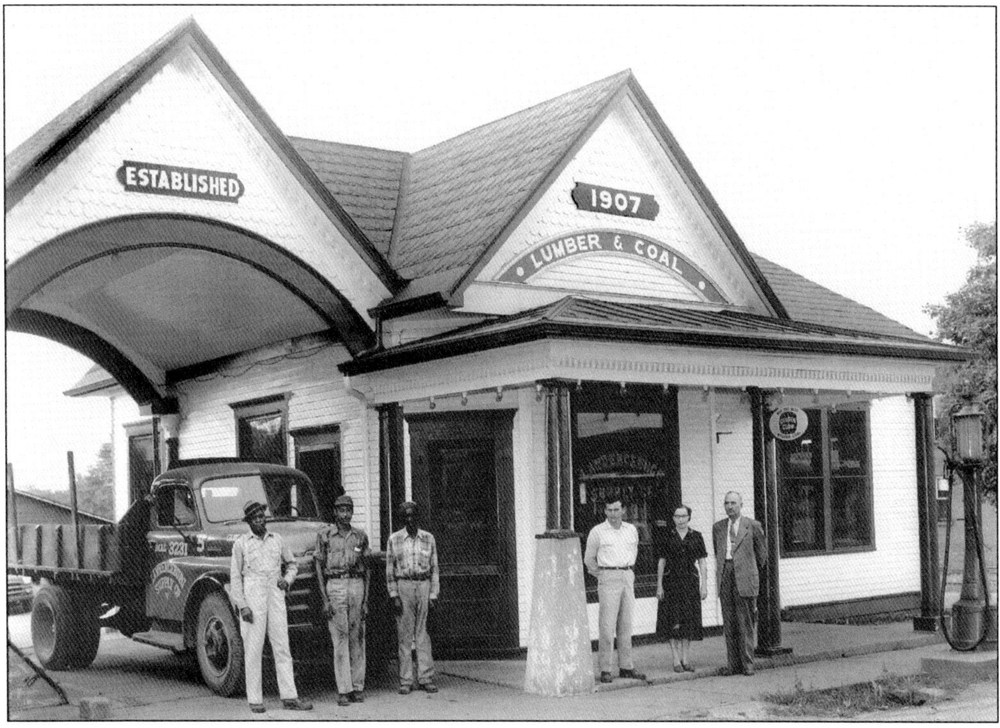

In 1907, W.A. Bottom founded the Lawrenceburg Supply Company, a business that mainly sold lumber. In 1925, Charles McGinnis purchased the company and ran it successfully until he sold it in 1964. The new owners added concrete and expanded the hardware portion of the business. Now, after over 100 years of continuous service, the Lawrenceburg Supply Company remains a family-owned enterprise successfully competing with the "big box stores." (Courtesy of Kitty McCoun.)

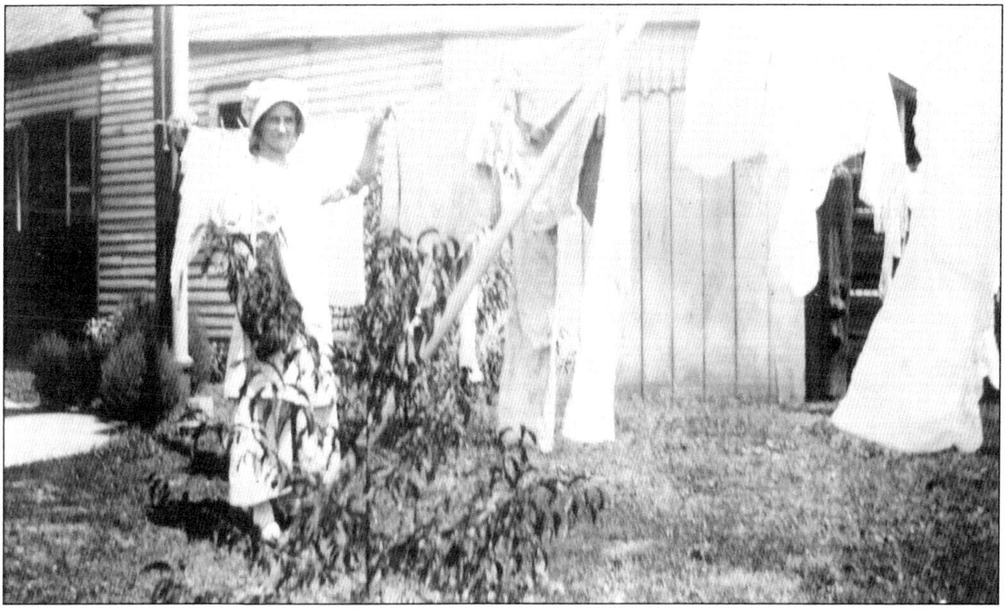

Wash day was an event for every household. Before electric clothes dryers, women hung their wash on an outside line and counted on the sun and wind to dry their clothes. (Courtesy of McFarland collection.)

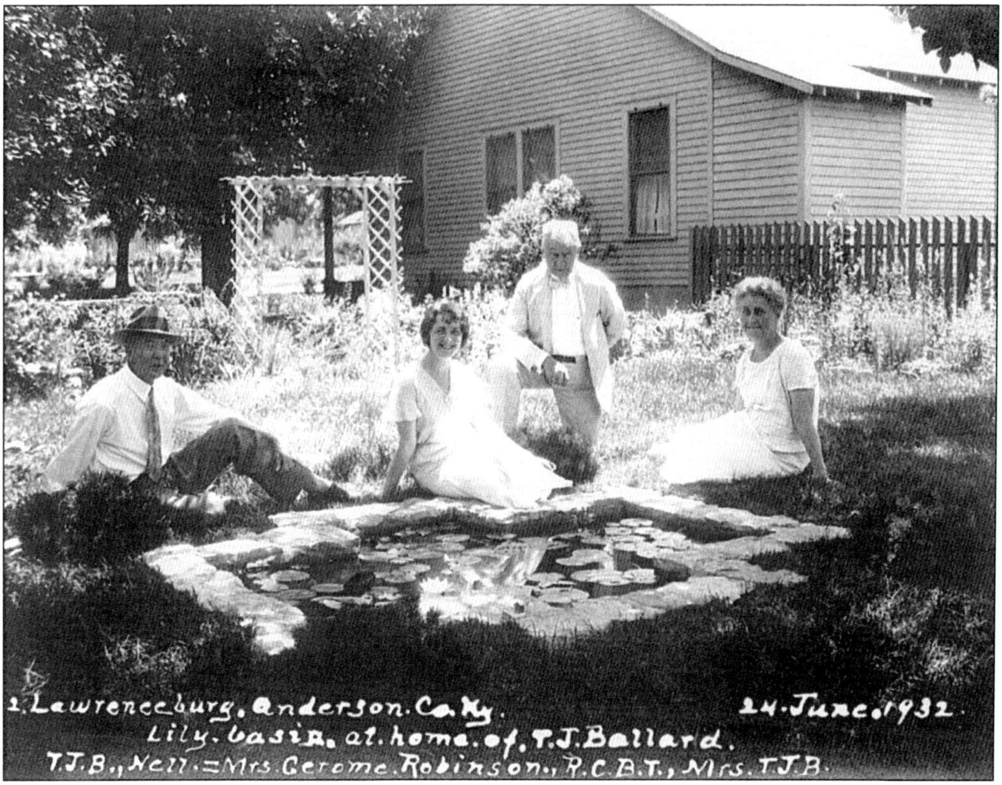

From left to right, T.J. Ballard, Nell Robinson, R.C. Ballard Thruston, and Gertrude Ballard relax by the lily basin in the side yard of the Ballard home in 1921. (Courtesy of Geoghegan collection.)

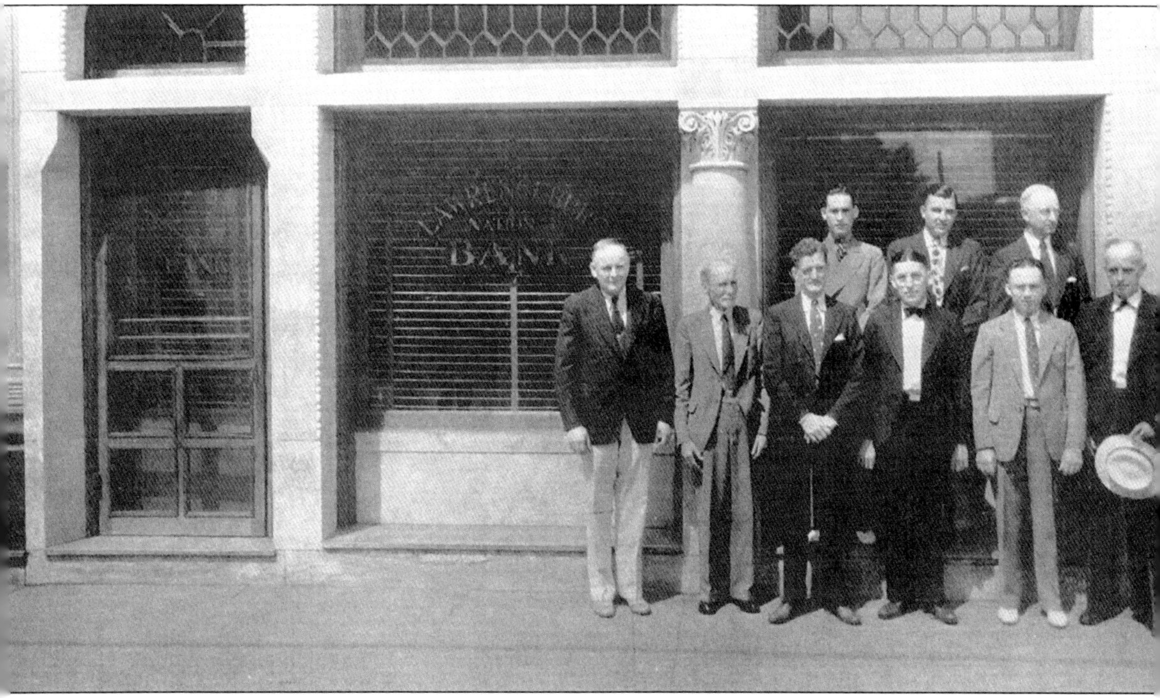

Lawrenceburg National Bank was one of Lawrenceburg's most stable institutions. The bank officials posing her are, from left to right, (first row) J.L. Sherwood, president; J.M.B. Birdwhistel, vice president; A.B. McAfee, chairman of the board; Ollie Calvert, cashier; Julian Cole, bookkeeper; and J.C. Bond, director; (second row) J.I. Blakemore, bookkeeper; William McGurk, assistant cashier; and James Sherwood, director. (Courtesy of Anderson County Historical Society.)

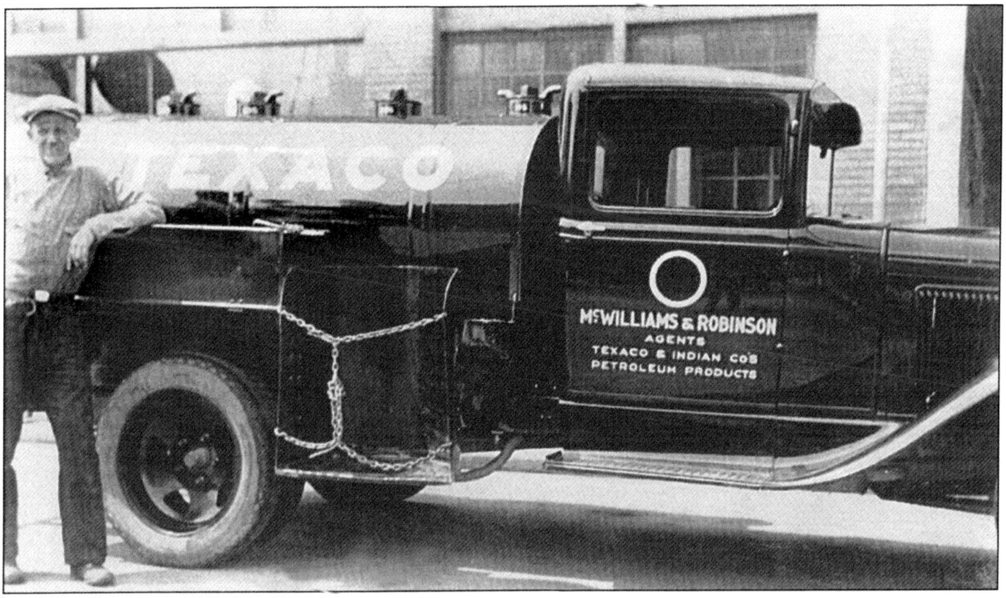

The Texaco Delivery Truck for the McWilliams and Robinson Distribution Company is seen here in front of the Robinson garage. By the 1920s, the automobile had replaced the horse as the main means of transportation. (Courtesy of Geoghegan collection.)

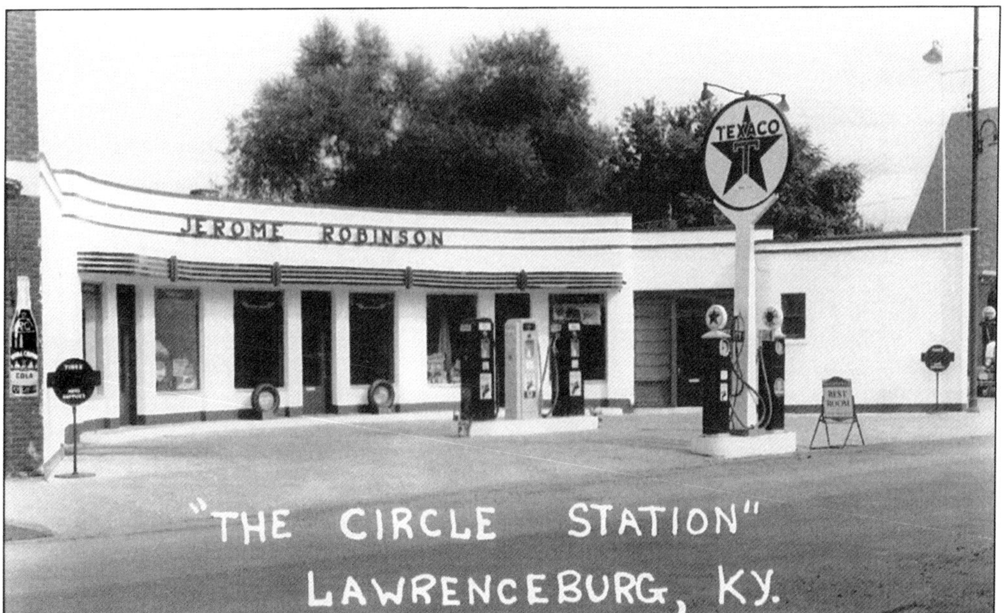

Jerome Robinson's Circle Station was a Texaco gas station on Main Street in downtown Lawrenceburg. (Courtesy of Geoghegan collection.)

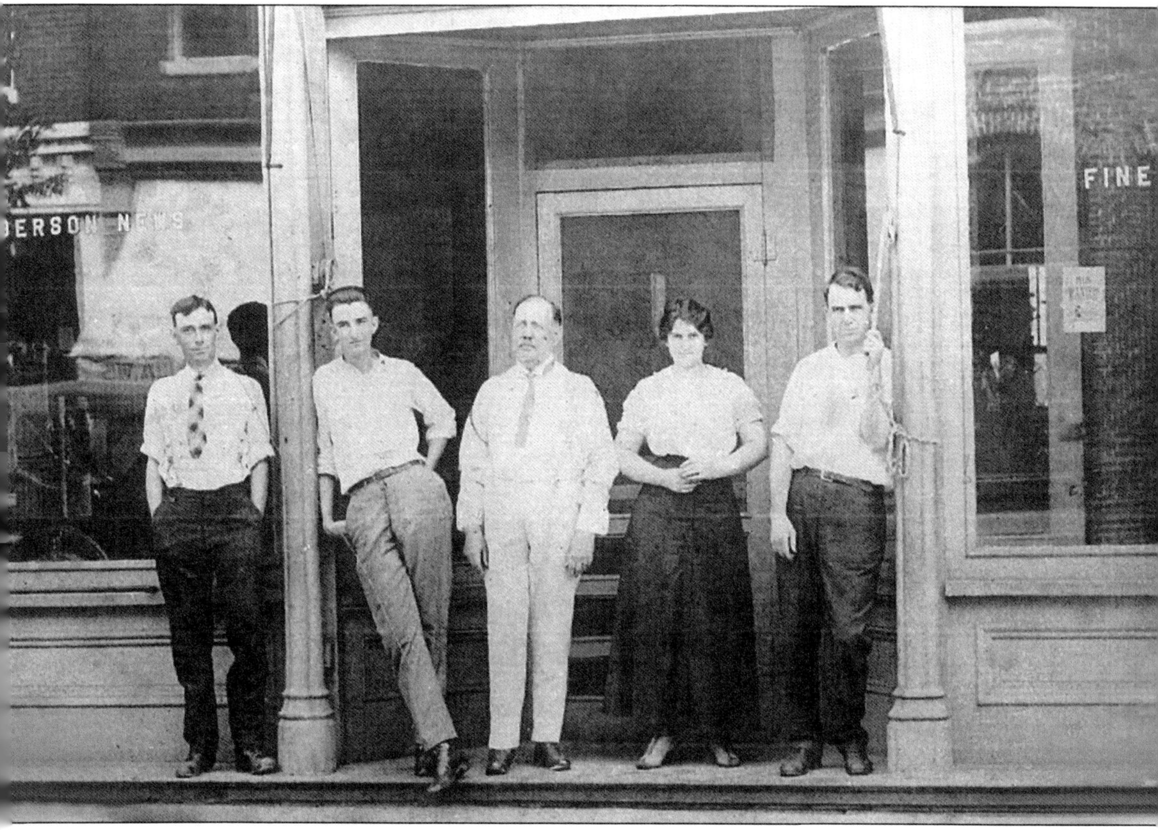

The *Anderson News* is the hometown newspaper, which is currently published on a weekly basis. At one time, Judge Faris Feland served as publisher and editor. Among those seen here are co-owner Elliott Garrison Sr. (far left) and foreman Morton Green (far right). (Courtesy of Anderson County Historical Society.)

The Anderson County Courthouse has always maintained a central position in downtown Lawrenceburg. It has burned on more than one occasion, first in 1859. The next courthouse, seen here, was built a year later of Kentucky River cliff stone and lasted until it burned in 1915. (Courtesy of Hawkins collection.)

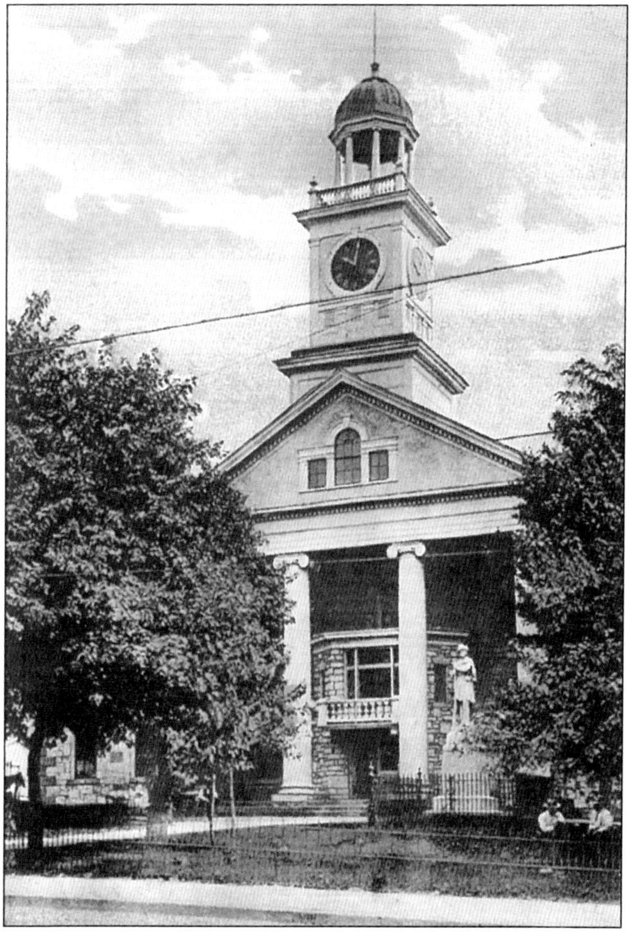

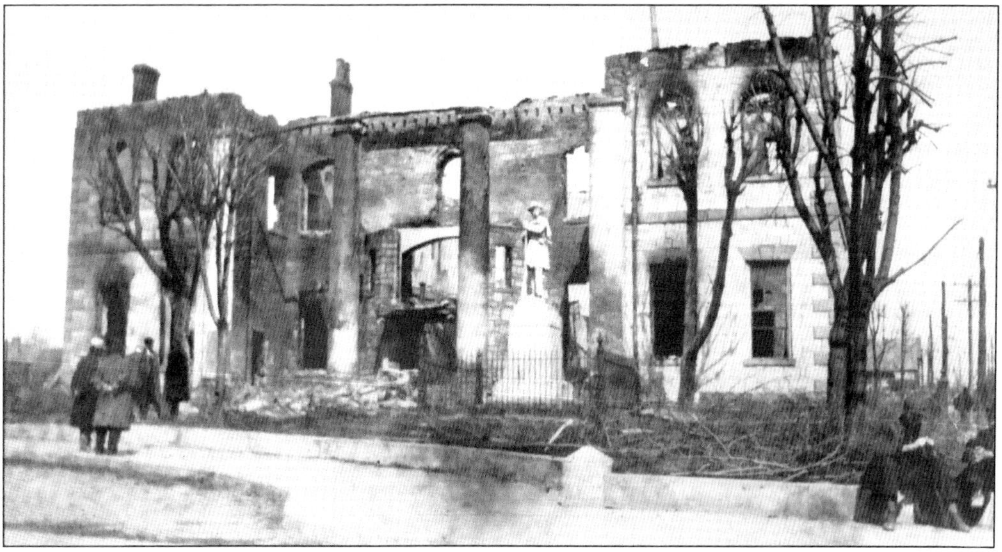

In 1915, fire destroyed the Anderson County Courthouse, leaving only these remains. (Courtesy of McAnly collection.)

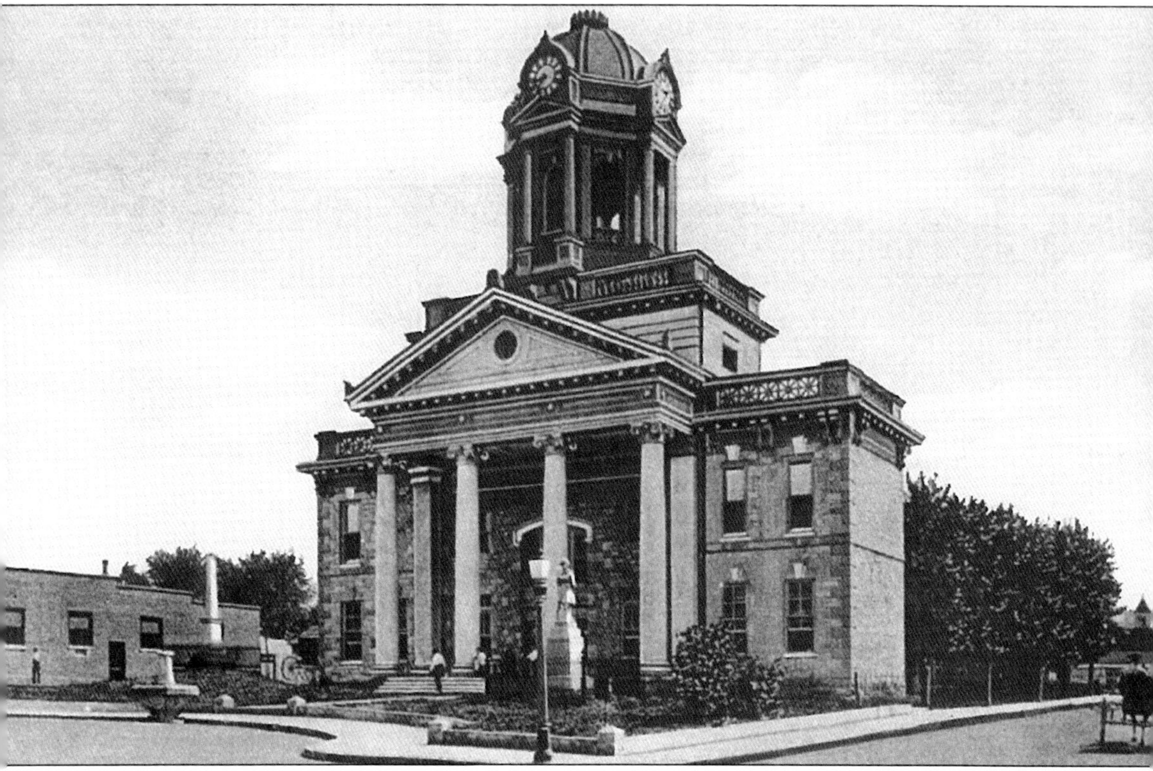

The present Anderson County Courthouse was built on the same location as the one that burned in 1915. Some differences are seen in the clock tower structure, but the overall design is similar to that of the previous building. (Courtesy of McAnly collection.)

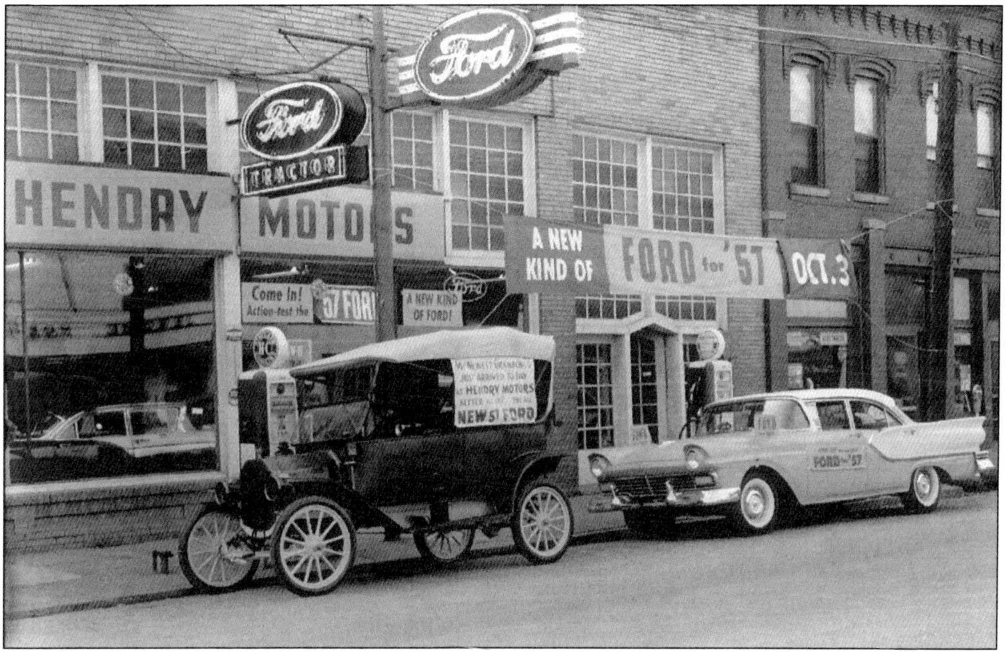

Hendry Motors was the Ford dealership in 1957, but it eventually closed and the building was razed. Today, the sheriff's offices are on this site. (Courtesy of Anderson County Historical Society.)

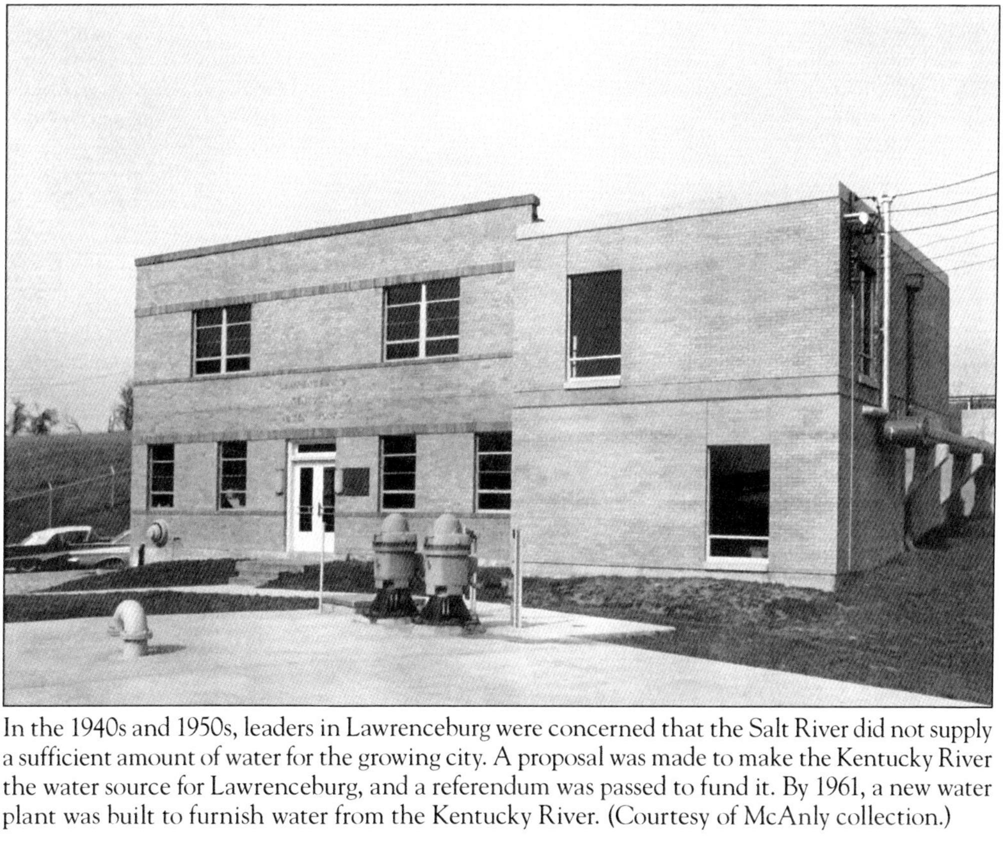

In the 1940s and 1950s, leaders in Lawrenceburg were concerned that the Salt River did not supply a sufficient amount of water for the growing city. A proposal was made to make the Kentucky River the water source for Lawrenceburg, and a referendum was passed to fund it. By 1961, a new water plant was built to furnish water from the Kentucky River. (Courtesy of McAnly collection.)

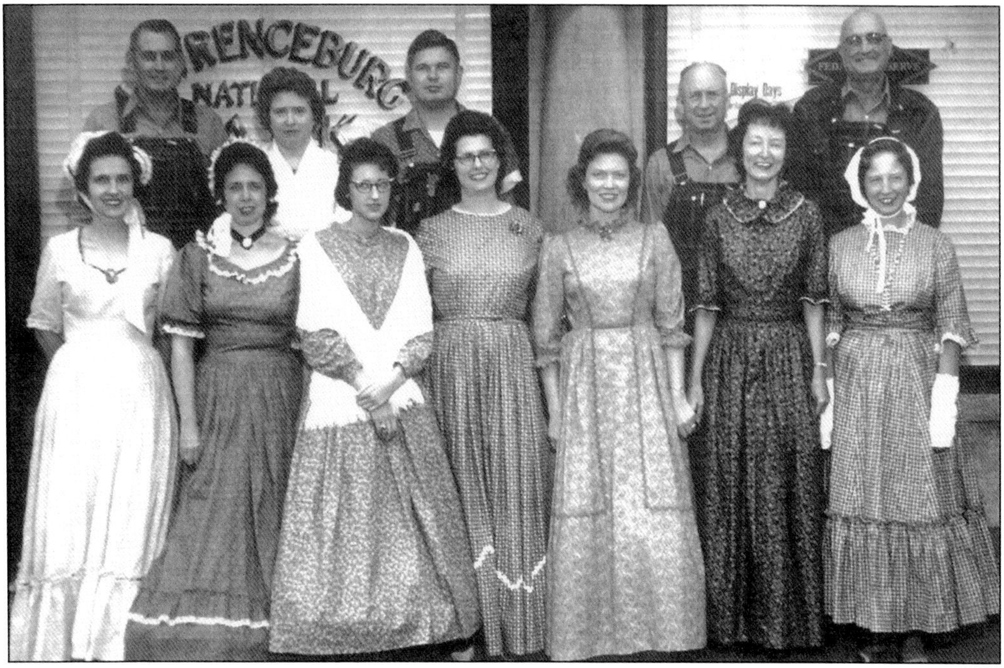

For several years in the 1960s, Lawrenceburg sponsored an event known as Hillbilly Days. Sponsors and participants were asked to dress in costume to reflect that event. Seen here from left to right at Hillbilly Days are Lawrenceburg National Bank staff members (first row) Frances Cox, Betty Young, Linda Hahn, Oneita Perry, Loretta Drury, Ruth Young, and Mabel Carey; (second row) Herman Morris, Hettie Simpson, Paul Vaughn Jr., Julian Cole, and Walter Major. (Courtesy of Anderson County Historical Society.)

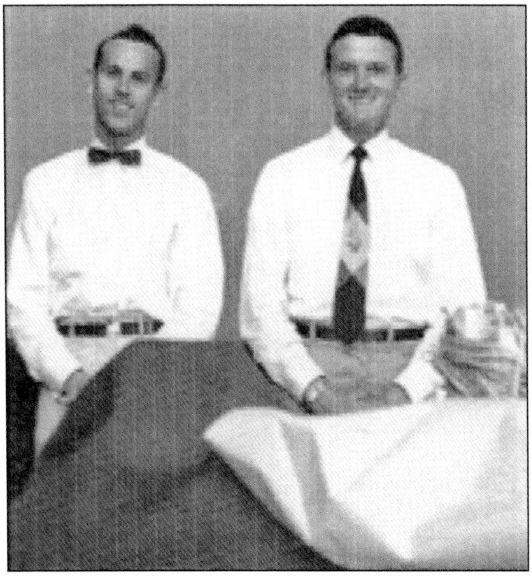

James Freeman (left) and Kenneth Mann maintained a partnership for more than 50 years in the Freeman and Mann Furniture Store in downtown Lawrenceburg. They were also both involved in civic and church events. (Courtesy of Tierney-Freeman collection.)

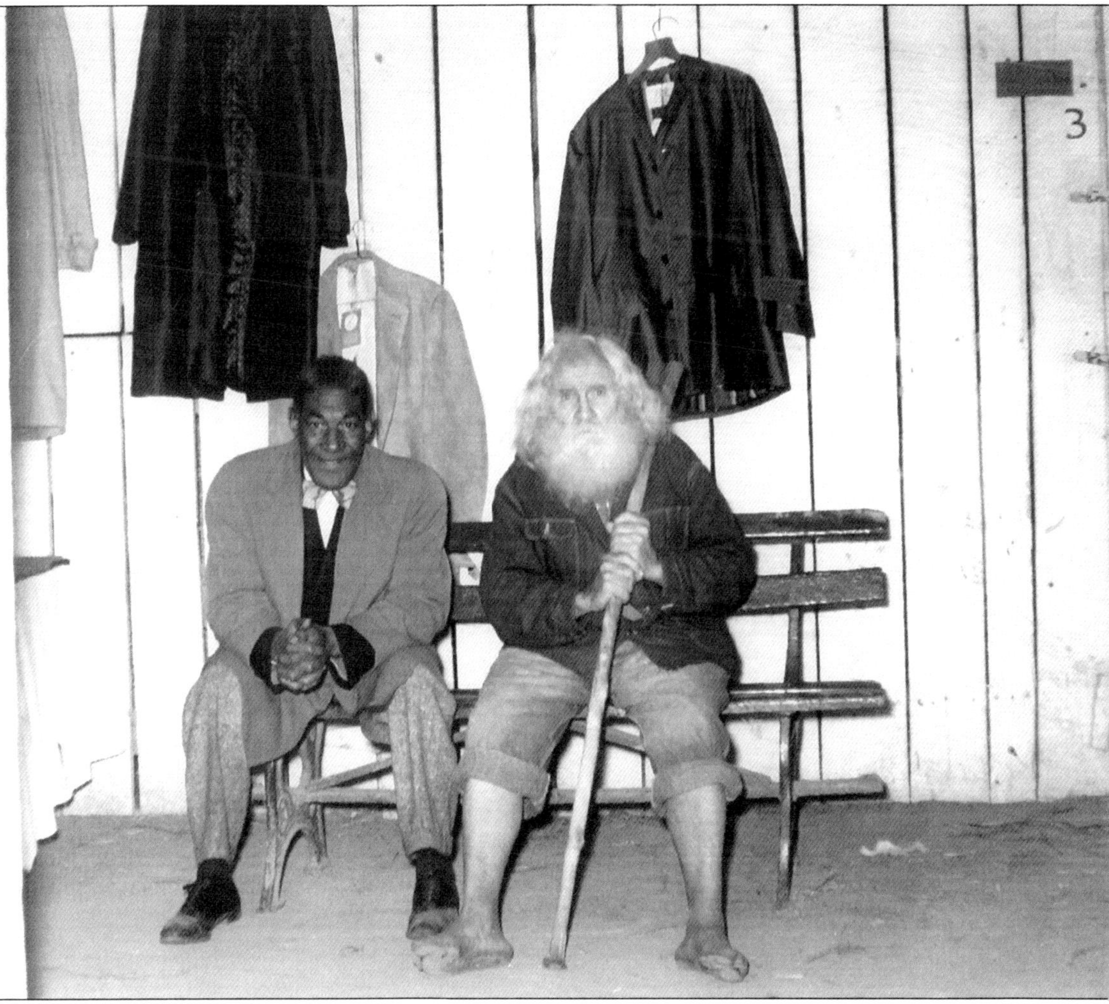

Both Harvey Blakeman and Ben Wilson made lasting impressions in Lawrenceburg. They are shown here at the county fair, where they were often seen. Both men were symbolic of Lawrenceburg: Ben with his staff, beard, and bare feet, and Harvey with his guitar and harmonica. (Courtesy of Smith collection.)

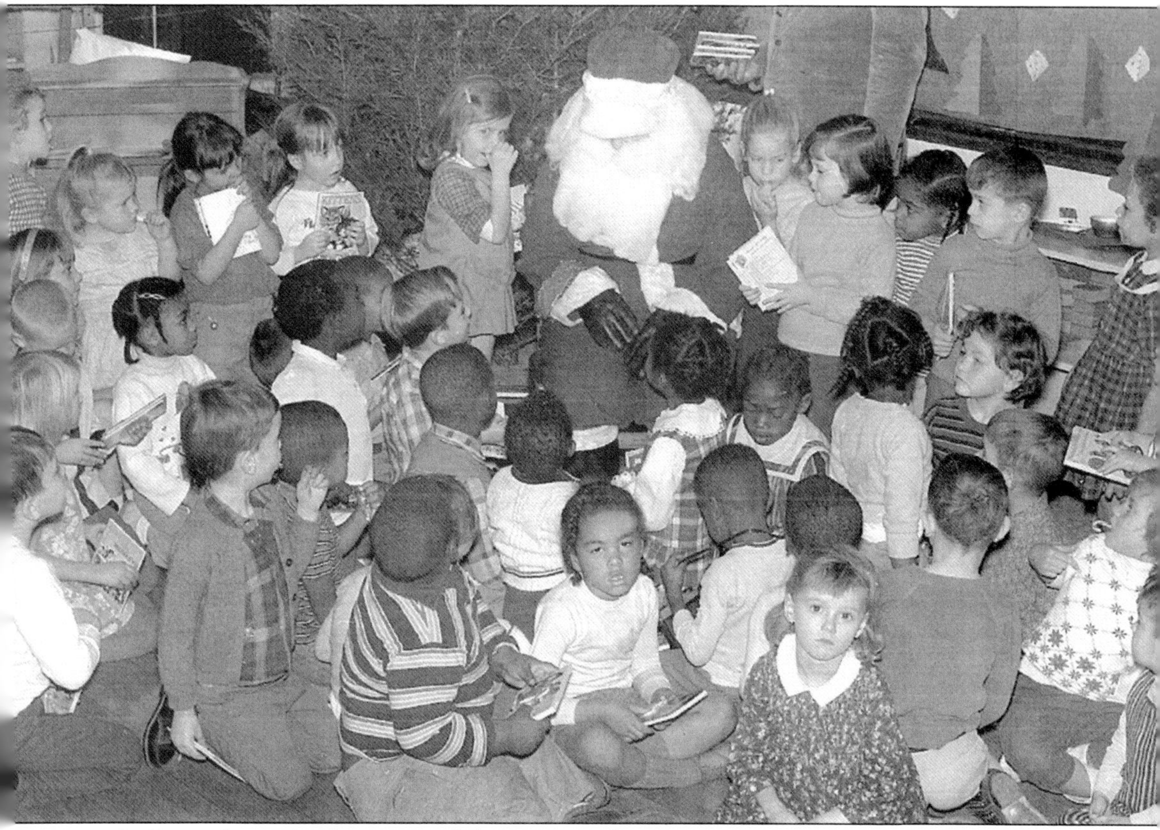

In December 1969, Santa visited the Lawrenceburg Community Center, where many children greeted him. Santa often gave the children gifts and candy during the holiday season. (Courtesy of Anderson County Historical Society.)

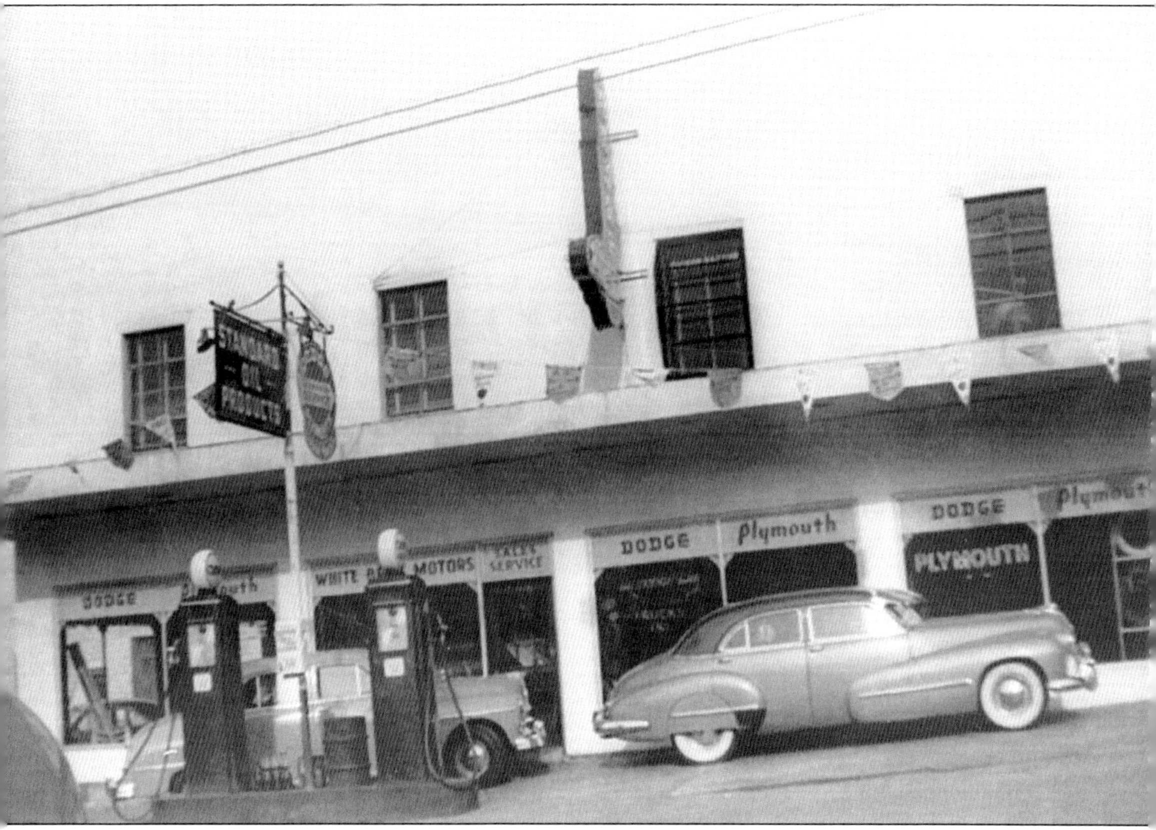

White Brothers Motors was a Dodge and Plymouth dealership in Lawrenceburg. J.R. White was involved with many businesses, including automobile lots, asphalt paving, car washes, and freight hauling. His brother Woodrow was usually in business with him. "Woodie" also tested his acting talents in the Hollywood film *The Flim-Flam Man*, which was filmed in Lawrenceburg. (Courtesy of McAnly collection.)

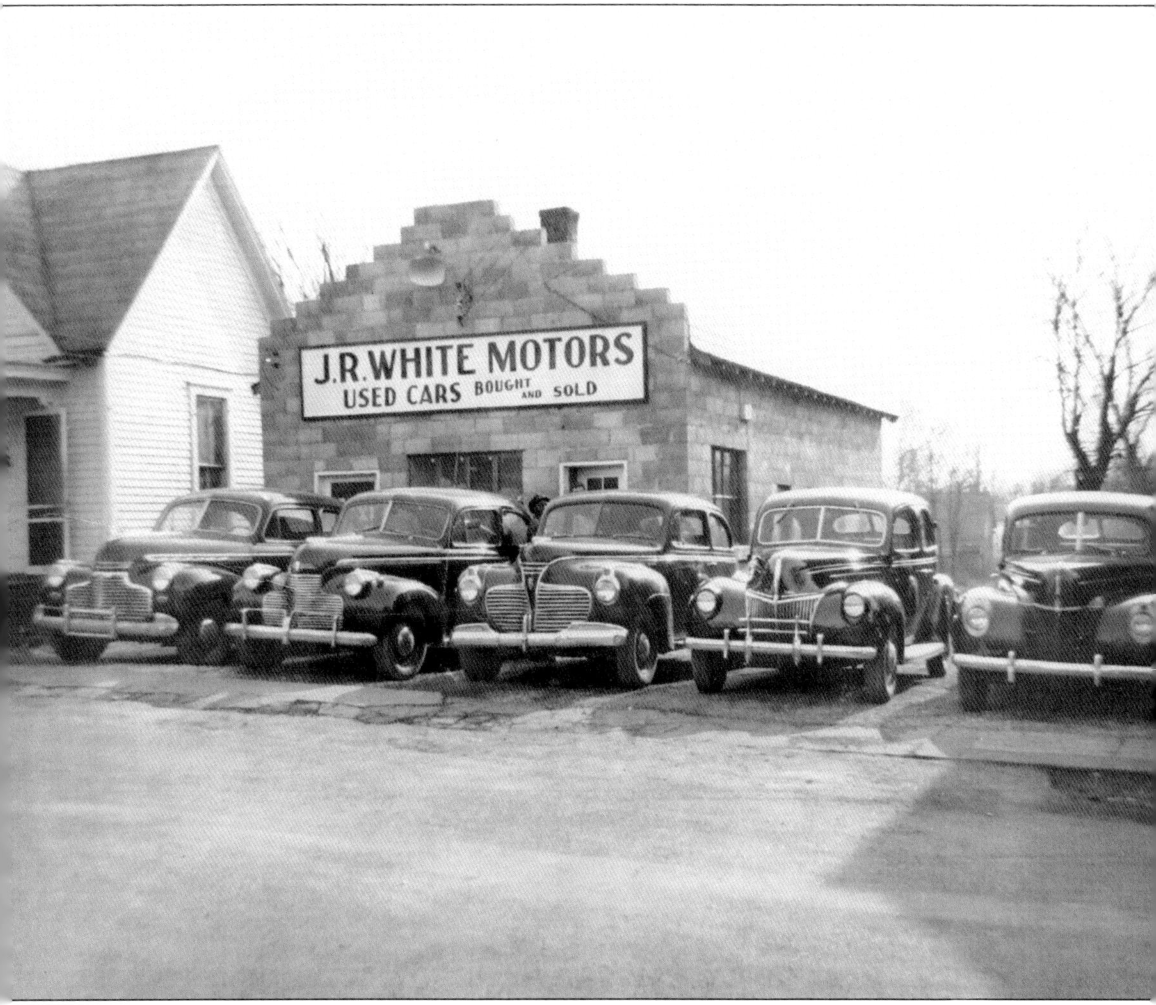

J.R. White Motors was on Court Street near some of White's other businesses. Unlike most automobiles today, black was the color of choice in the 1930s and 1940s—as it was usually the only one. (Courtesy of McAnly collection.)

Seven

Recreation and Entertainment

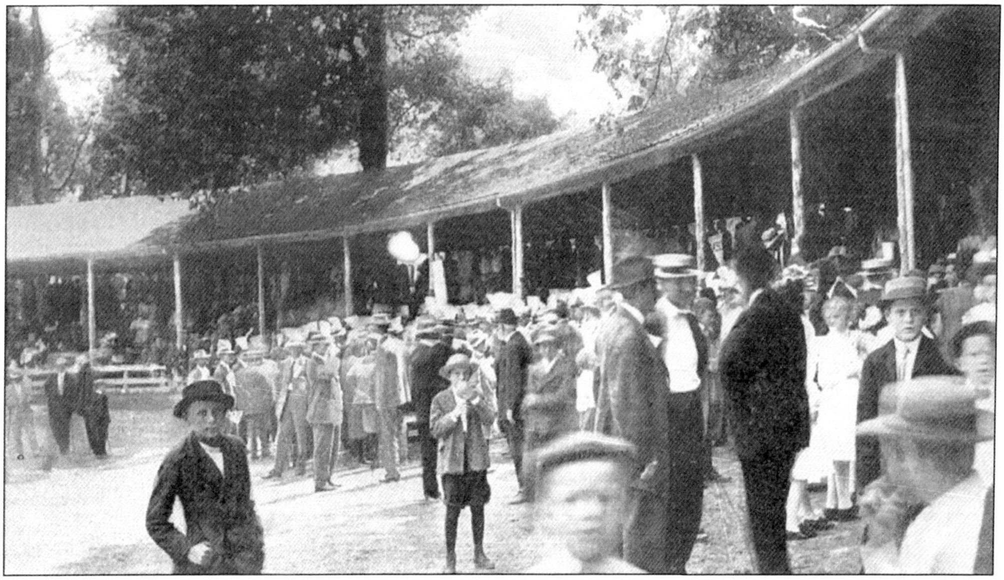

The Lawrenceburg Fair and Horse Show was held in August 1909 in the McKee Woodlands in the vicinity of Carlton Lane. The fair was eventually purchased by the American Legion and moved to its present site on Highway 62. (Courtesy of Hawkins collection.)

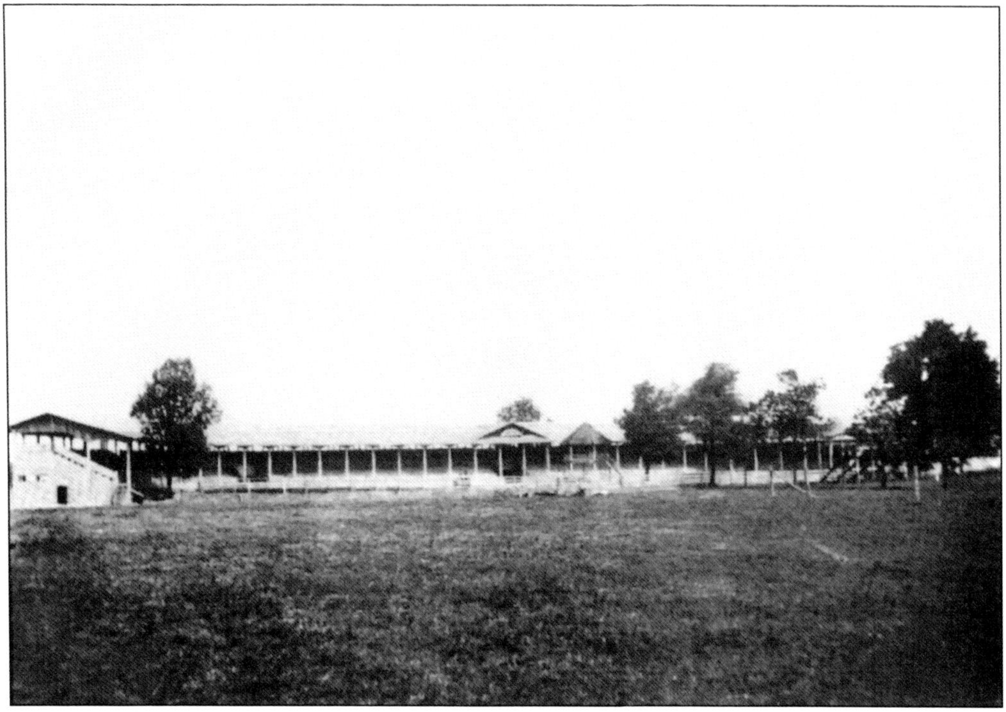

This 1915 photograph of the fairgrounds on the McKee property show the grandstand that was privately built. These fairgrounds were Lawrenceburg's first, although the first fair started near Salvisa. (Courtesy of Hawkins collection.)

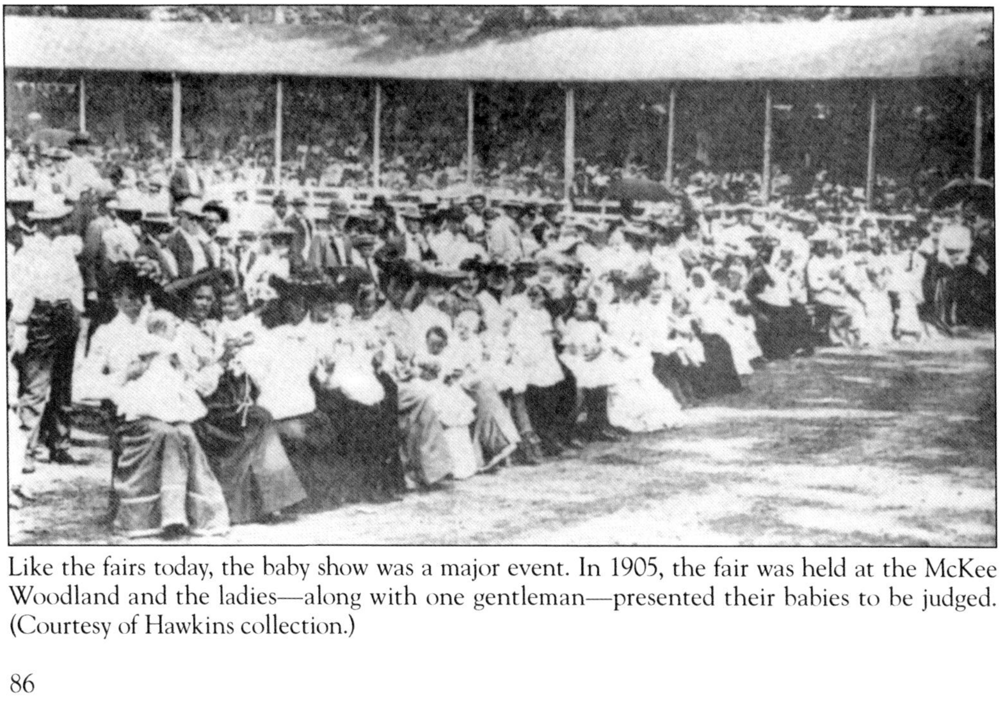

Like the fairs today, the baby show was a major event. In 1905, the fair was held at the McKee Woodland and the ladies—along with one gentleman—presented their babies to be judged. (Courtesy of Hawkins collection.)

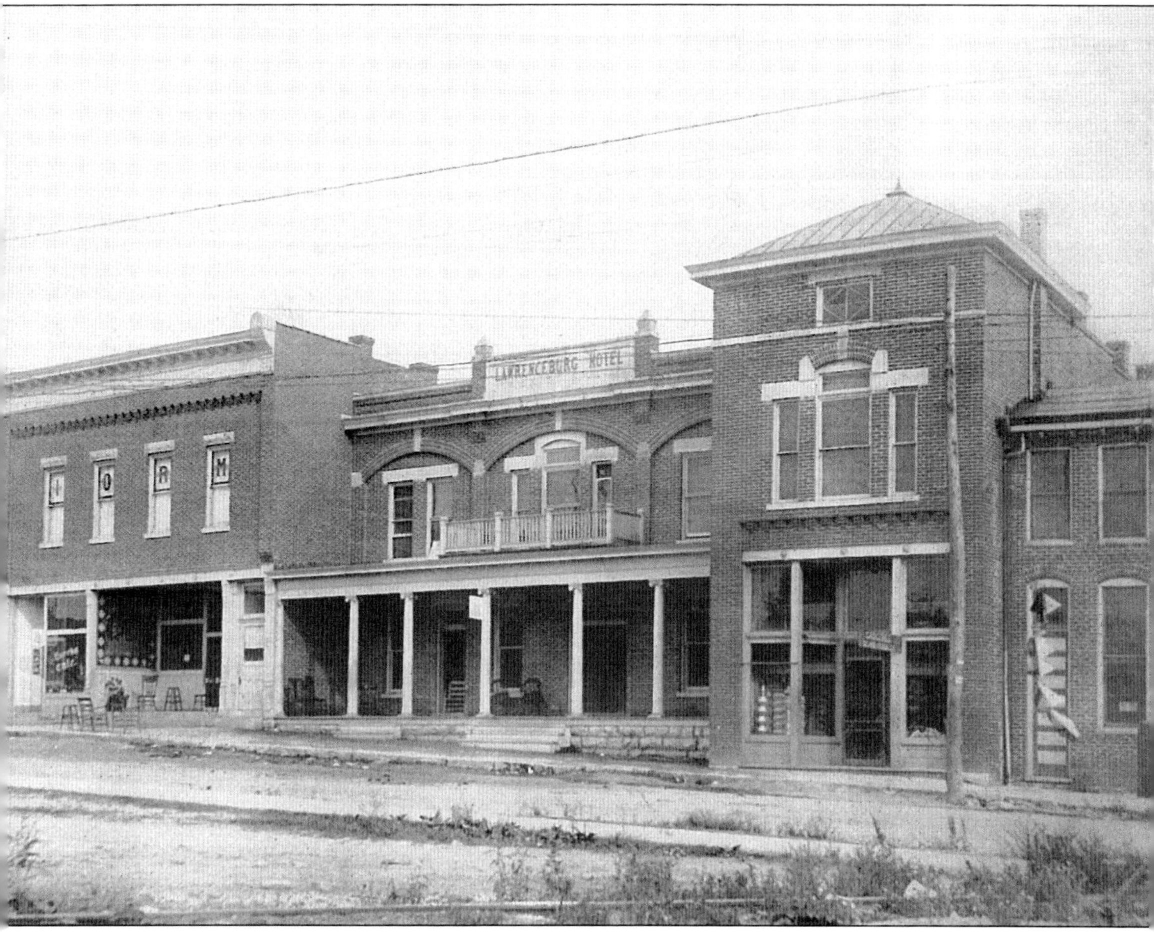

The Lawrenceburg Hotel, originally known as the Wabott Hotel, was erected in 1891–1892 by W.A. Bottom at a cost of $37,000. The hotel, along with the opera house (far right), provided comfortable rooms and entertainment for travelers and local inhabitants. The hotel building also included a small cafe. The hotel was located on Court Street across from the depot. In 1906, it advertised that its cuisine was unexcelled, it had electric lights and waterworks, and everything was clean. Rates were listed at $2 per day. (Courtesy of McAnly collection.)

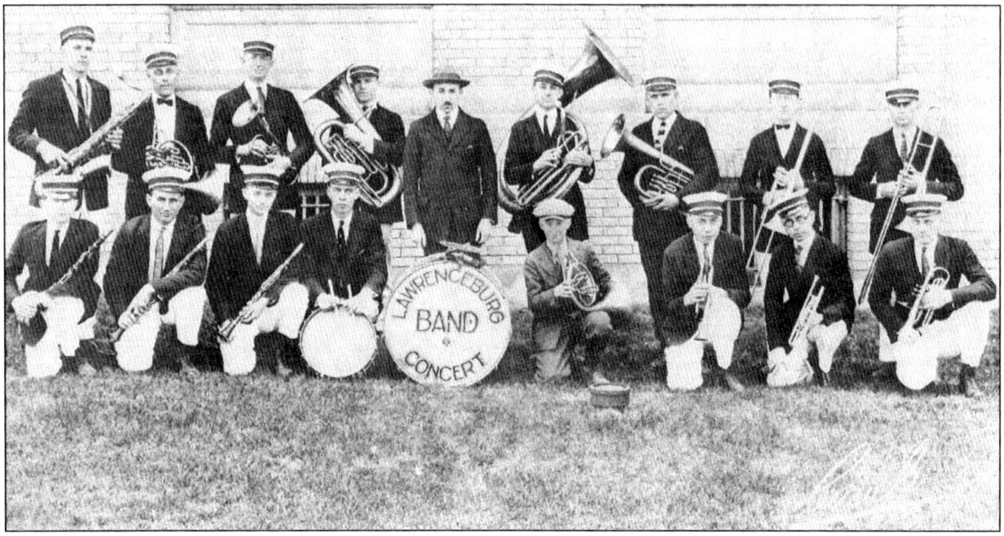

Lawrenceburg's concert band in 1926 was directed by Professor A.J. Giancola. Band members included, from left to right, (first row) Marvin Wash, Allen Wash, Howard Garner Jr., Carl Duncan, Eddie J. Morris, Ballard Hawkins, Harry Brunk, and Frank Martin; (second row) Jim Bailey McGurk, Al Jacobs, Stanley Johnson, Goebel Crossfield, Professor Giancola, Howard Garner Sr., Luther Satterwhite, Robert Mays, and Julian Cole. (Courtesy of Jack Morris.)

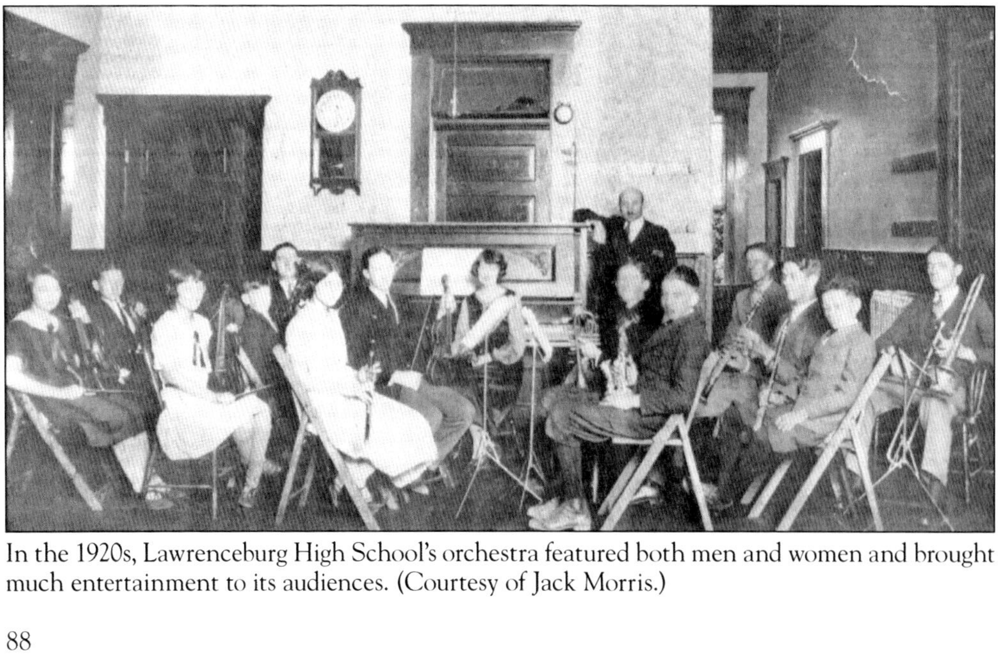

In the 1920s, Lawrenceburg High School's orchestra featured both men and women and brought much entertainment to its audiences. (Courtesy of Jack Morris.)

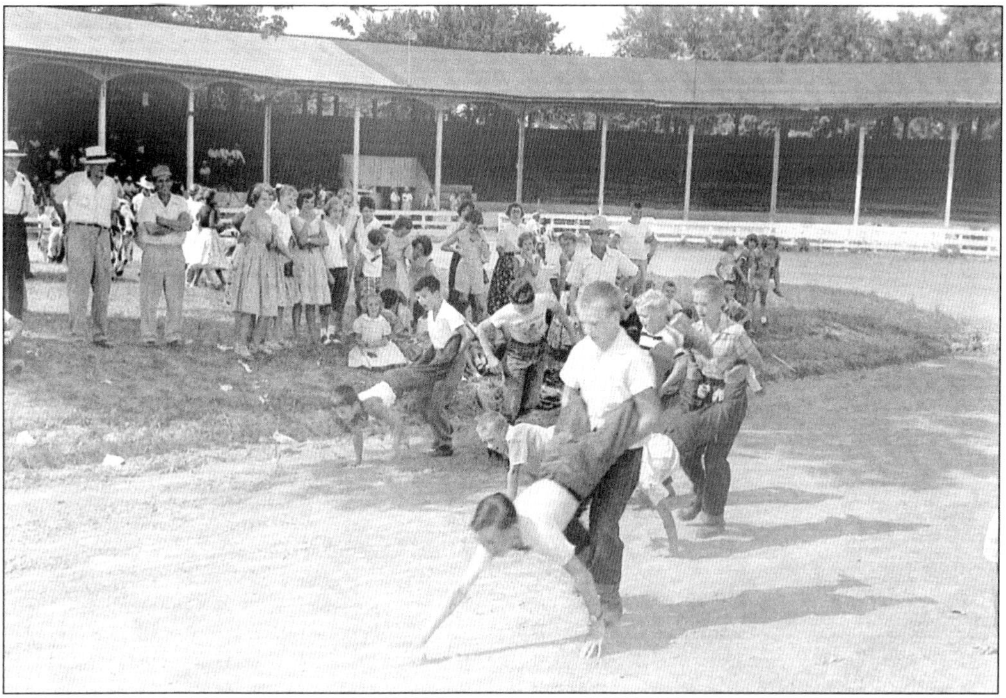

Wheelbarrow races were popular for elementary students at the fair. (Courtesy of Hanks collection.)

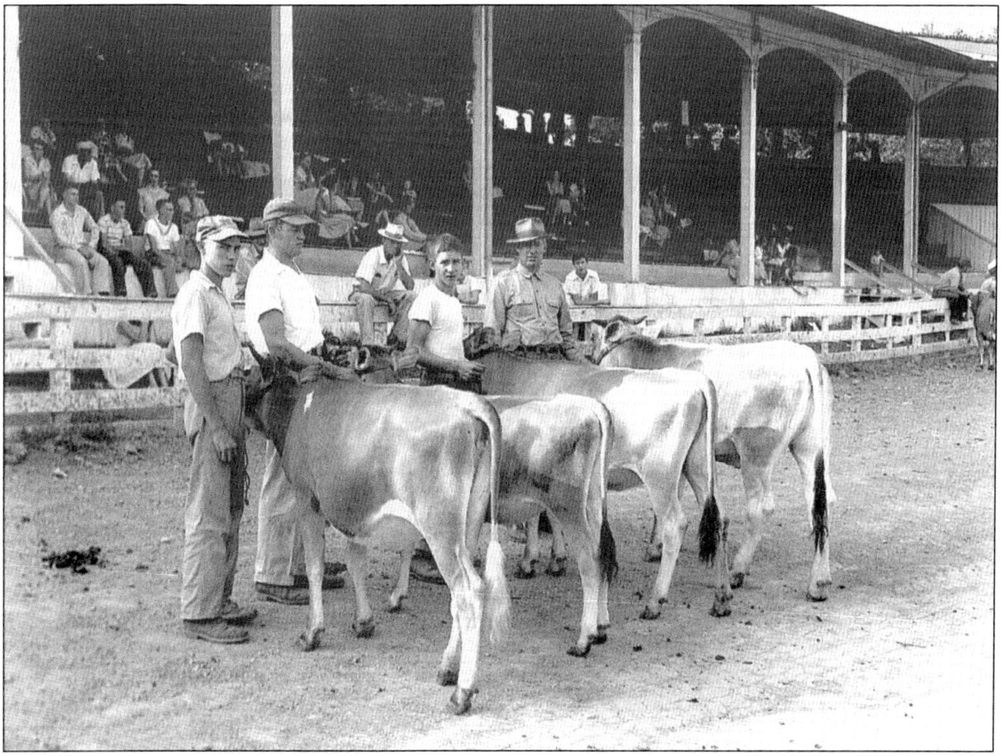

Young farm boys show their dairy cattle at the fair. The cows were judged and ribbons were given for placement. (Courtesy of Hanks collection.)

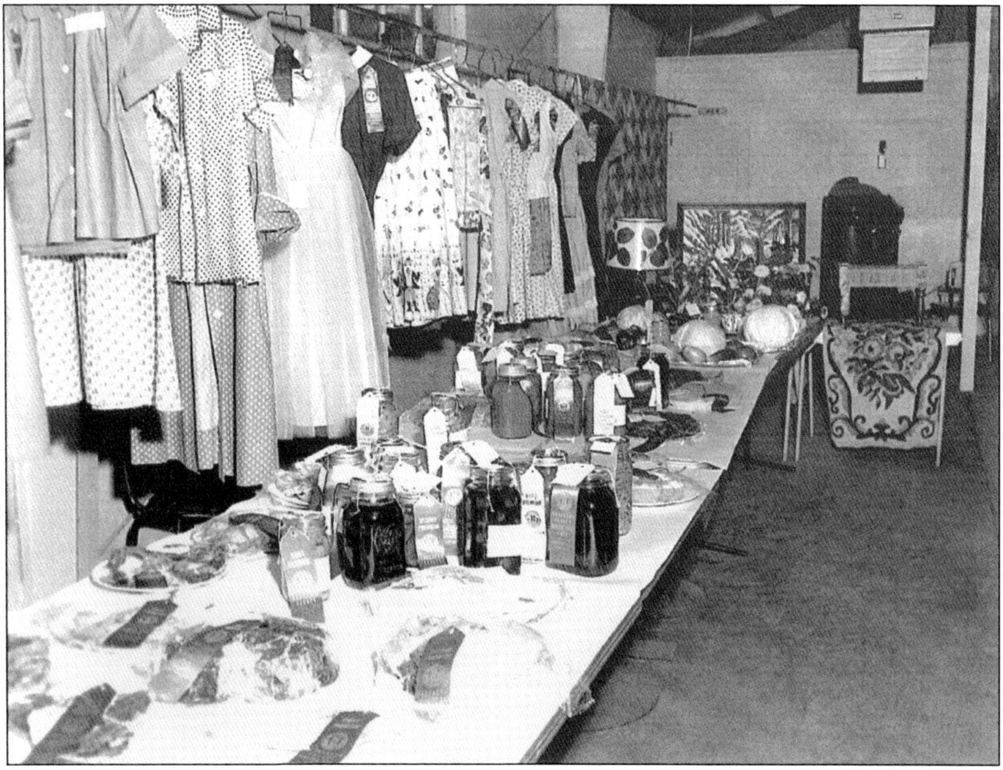

Ribbons indicate the prizewinners in the categories for preserves, canned fruit, vegetables, sewing, and other household products at the fair. Each year, the competition was great, and local winners moved onto the state fair in Louisville. (Courtesy of Hanks collection.)

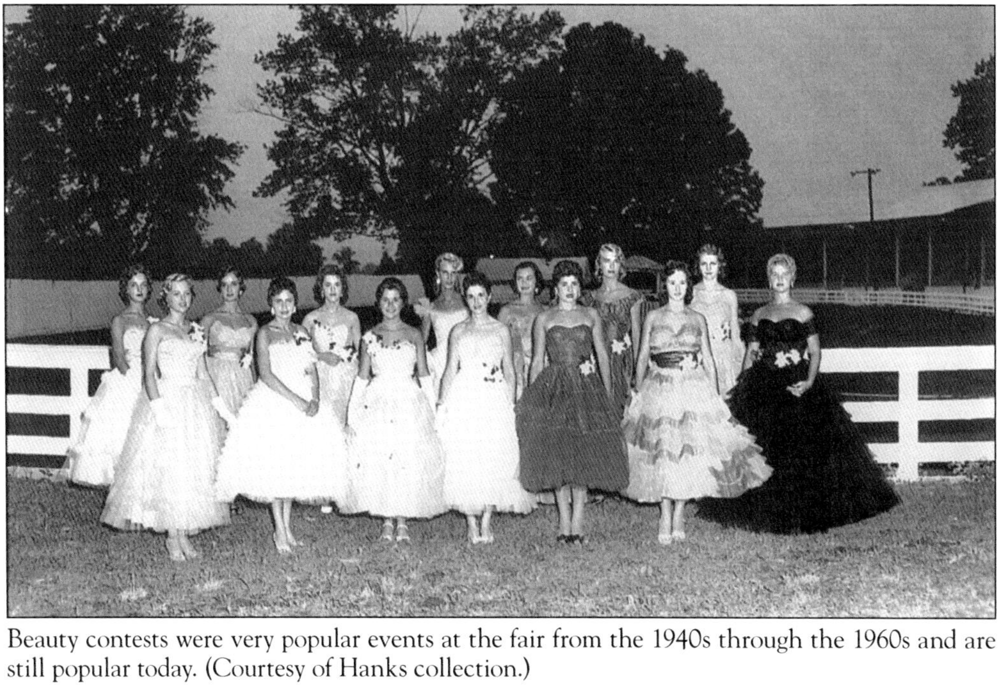

Beauty contests were very popular events at the fair from the 1940s through the 1960s and are still popular today. (Courtesy of Hanks collection.)

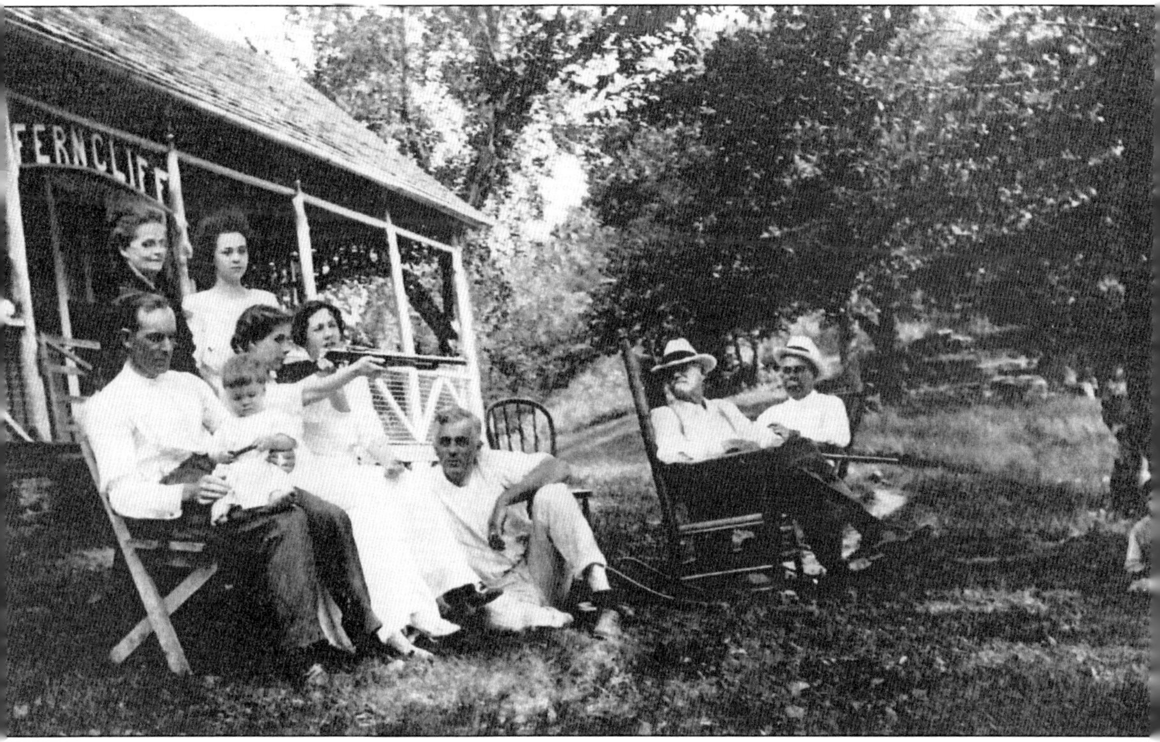

Camps on the Kentucky River provided sites for weekend and summer relaxation. In the summer of 1914, the Ripy family met at their camp, Fern Cliff. The family included, from left to right, (first row) E.W. Ripy Sr. (holding E.W. Ripy Jr.), Madeline Johnson Ripy, unidentified, Matt Searcy, Capt. Polk Searcy, and J.M. Johnson; (second row) Sallie Johnson and Mary Searcy. (Courtesy of Ripy-Crawford collection.)

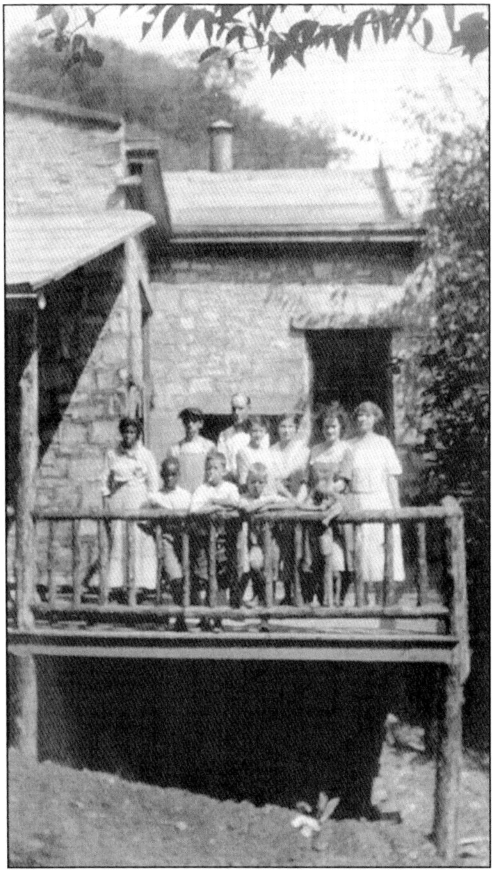

Bat Rock was converted from a former Cedar Brook Distillery pump house into a river camp. The camp was used in the 1920s and 1930s, but it later fell into disrepair after Kentucky Utilities built a coal-fired, electric-generated plant directly across the river from it. (Courtesy of Ripy-Crawford collection.)

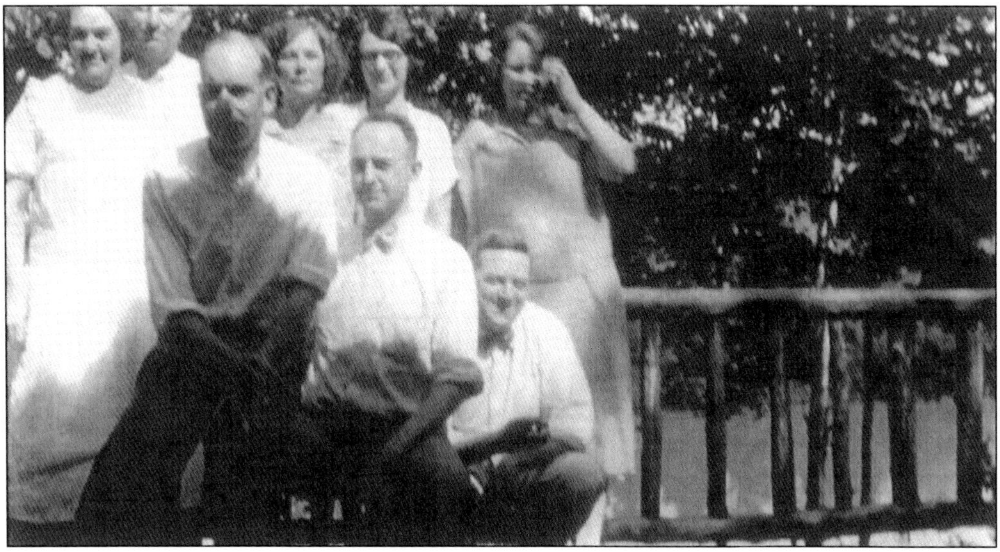

Guests at Bat Rock seen here in 1922 include Keen Johnson (center, in the front row of men), the co-owner and publisher of the *Anderson News*. In 1925, he purchased a paper in Richmond, Kentucky, and moved away from Lawrenceburg. He later became Kentucky's 45th governor, serving from 1939 to 1943. (Courtesy of Ripy-Crawford collection.)

Trap shoots were held for many years, especially in the 1940s and 1950s. Lin Mountjoy is pictured around 1925 taking aim at a clay pigeon. (Courtesy of Mountjoy collection.)

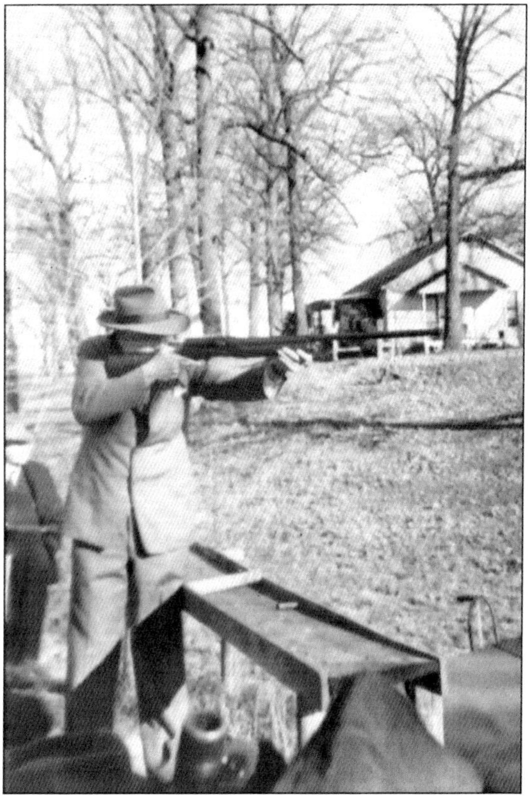

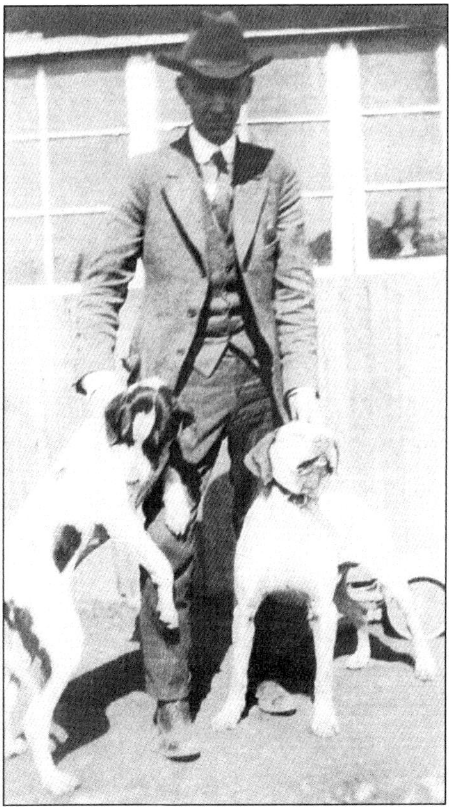

Hunting dogs were prized by their masters in the area. William Ripy is seen here around 1938 handling his favorite bird dogs, which were used in hunts and field trials. (Courtesy of Geoghegan-Pike collection.)

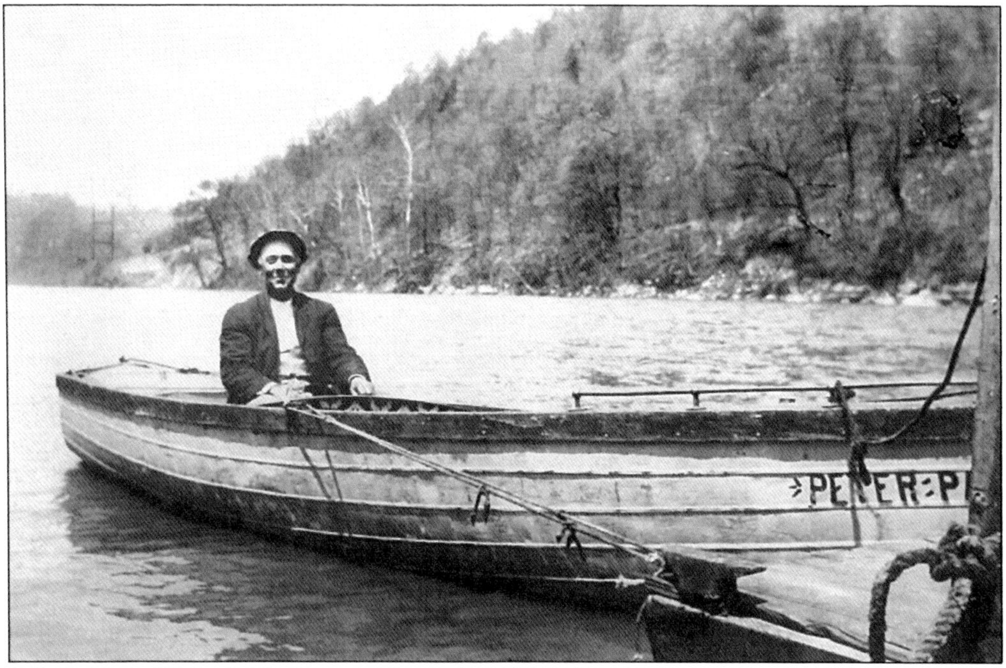

Fishing in the Kentucky River near Tyrone has always been a popular sport. Trotlines were used by locals, but game fishermen preferred poles and live bait. (Courtesy of Hawkins collection.)

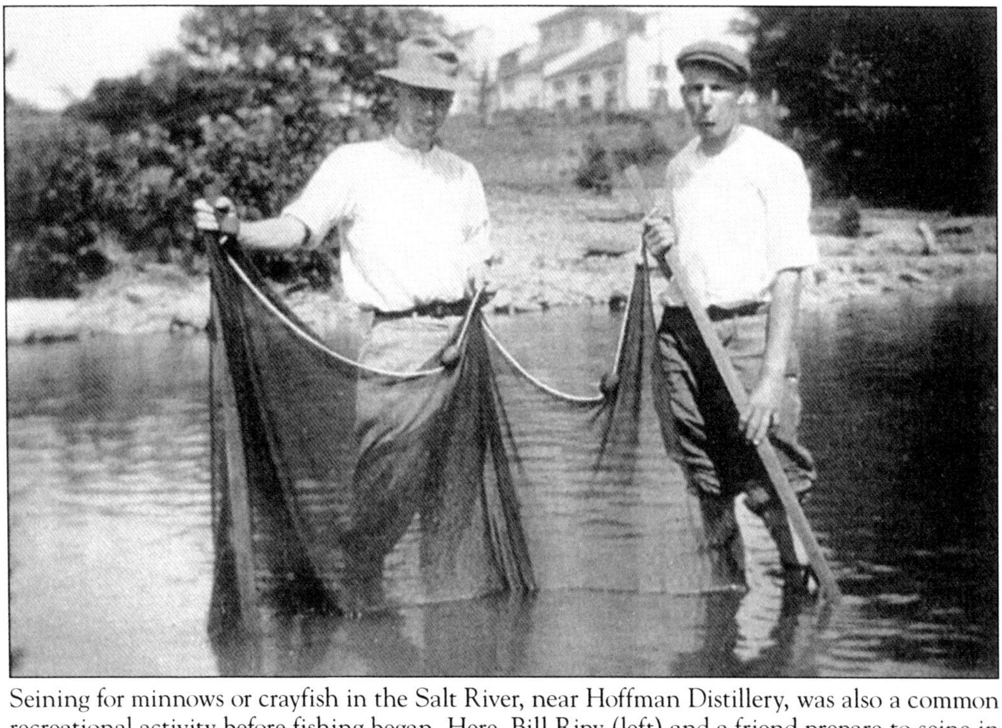

Seining for minnows or crayfish in the Salt River, near Hoffman Distillery, was also a common recreational activity before fishing began. Here, Bill Ripy (left) and a friend prepare to seine in the shallows of the river. (Courtesy of Geoghegan-Pike collection.)

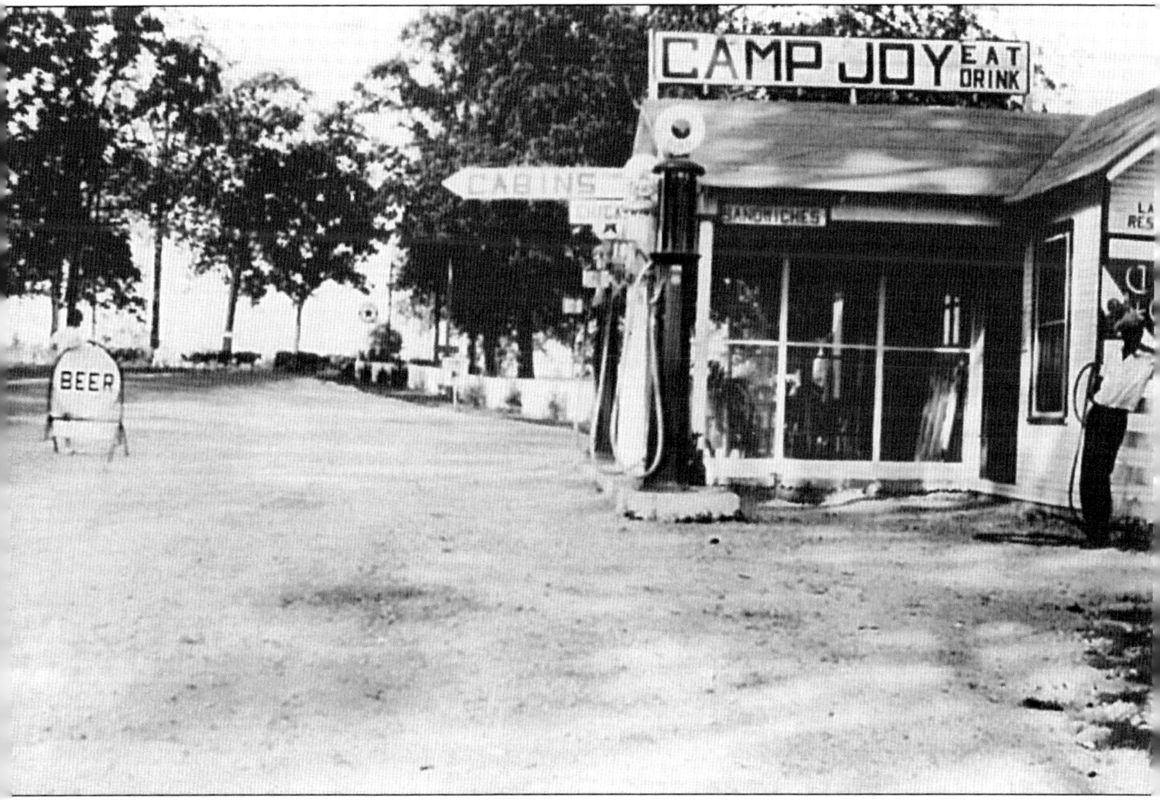

Camp Joy, on the Mountjoy farm, had rental cabins as early as the late 1920s, long before motels became popular. The office at Camp Joy was also a place to purchase food, fuel, and drinks and rent cabins and horses. Later, the office was converted to an appliance store, which sold electric milkers and large household appliances. (Courtesy of Mountjoy collection.)

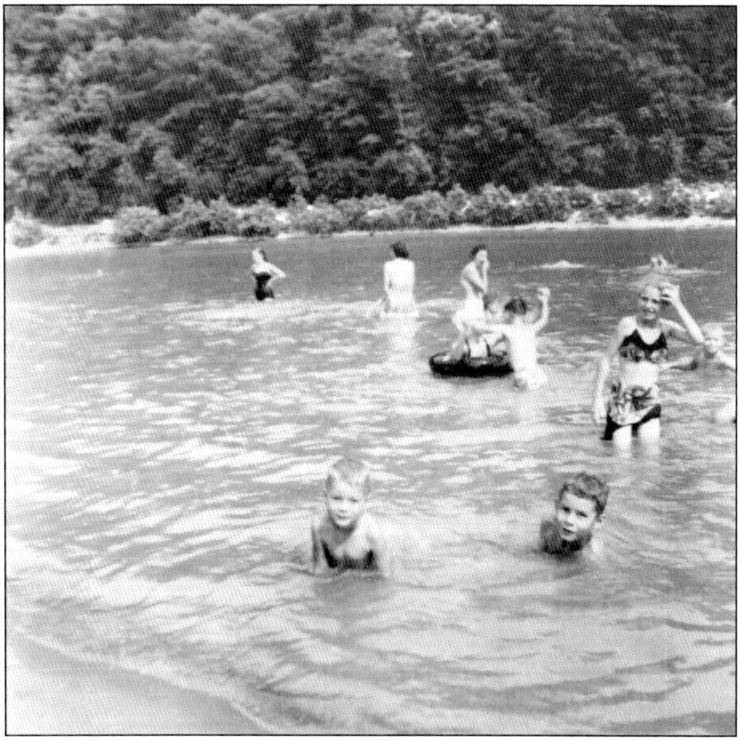

Swimming and boating at Lock Five on the Kentucky River were popular weekend activities in the 1940s and 1950s. The polio epidemic in the early 1950s frightened many people into cutting down on activities in natural waters. Here, in front, Bobby (left) and Billy Bryant swim with others at Lock Five. (Courtesy of Bryant collection.)

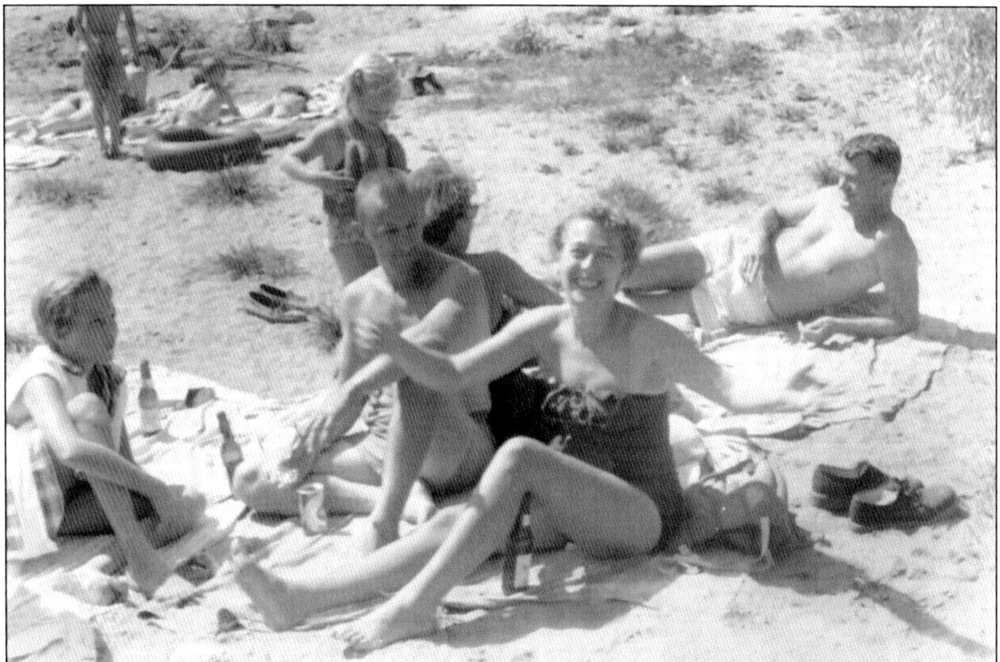

The Mountjoy family enjoys an afternoon of sunbathing and relaxation on the beach at Lock Five around 1952. (Courtesy of Mountjoy collection.)

Houseboats were favorite vessels to navigate and enjoy the Kentucky River. Many families shared leisure times and memories of these excursions. (Courtesy of Mountjoy collection.)

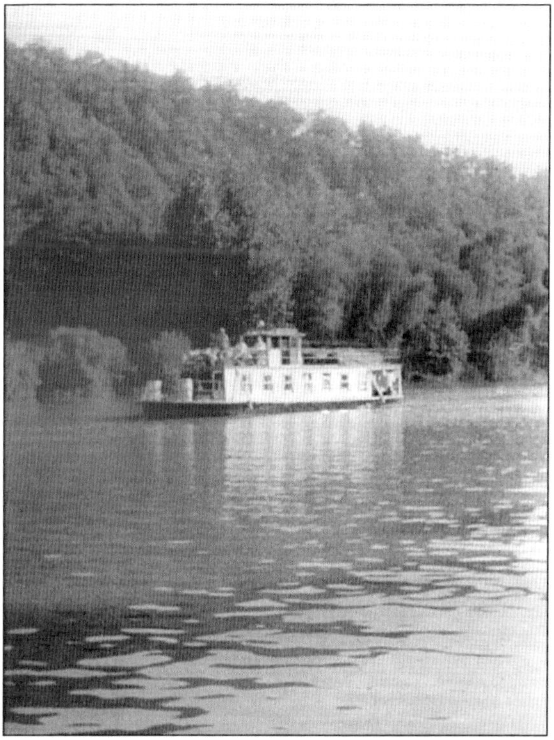

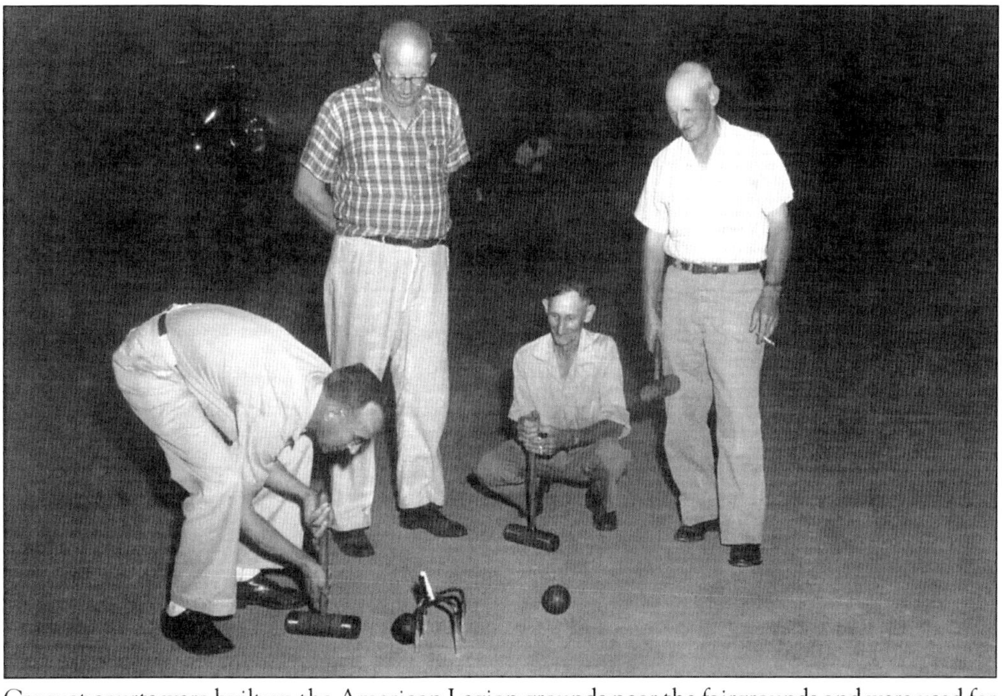

Croquet courts were built on the American Legion grounds near the fairgrounds and were used for a number of years. The Kentucky State Croquet Tournaments were held here in the early 1960s. Silas Dennis (far left) and Bill Nevins (second from left) teamed up to defeat their opponents in 1960. (Courtesy of J. Ripy collection.)

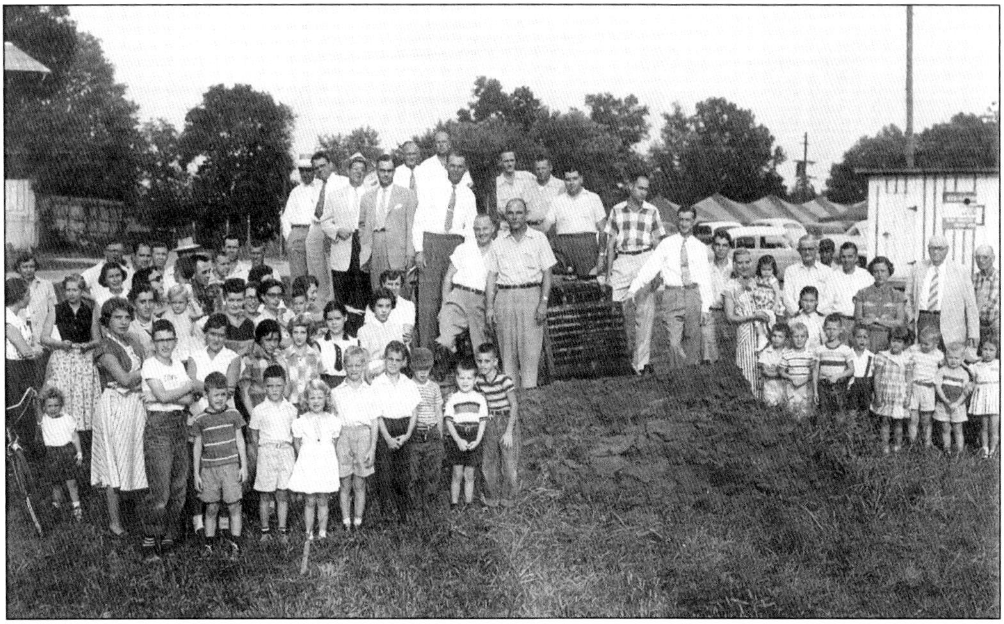

In 1954, a ground-breaking ceremony was held at the American Legion property for a new community swimming pool. Officials and townspeople gathered as a bulldozer broke the ground. In the center of this photograph (with the white short-sleeve shirt) is Lawrenceburg mayor Tom Ripy, who was instrumental in this community project. The pool was not dedicated until May 30, 1957. (Courtesy of J. Ripy collection.)

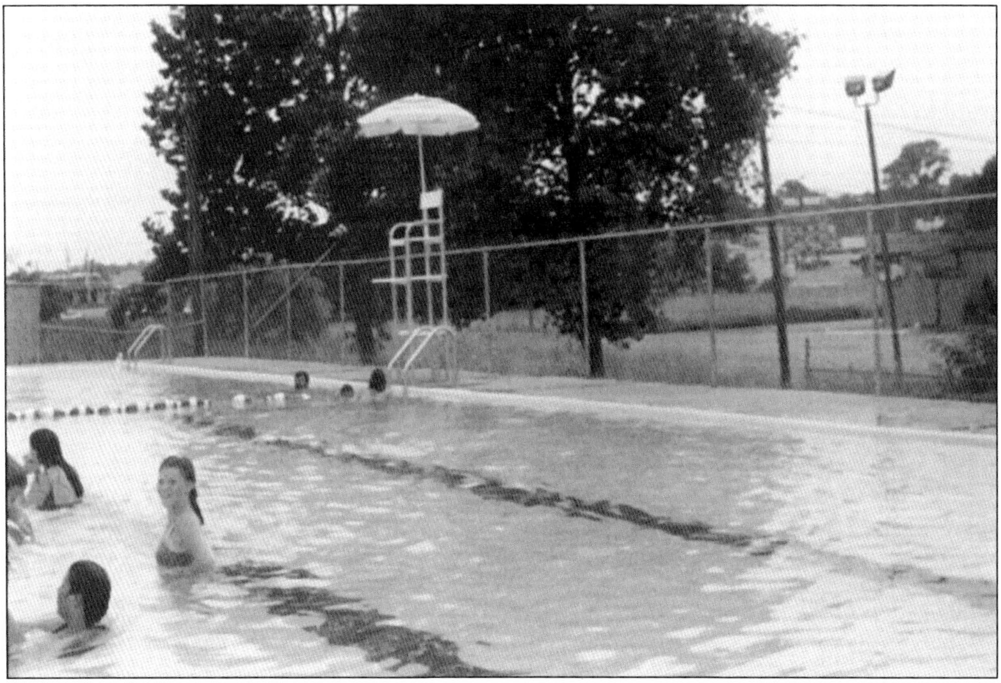

The pool had been operating for several years when this photograph was taken. It was managed for several years by one of the high school teachers or a college student home for the summer. The pool was often quite crowded, especially on weekends. (Courtesy of Green collection.)

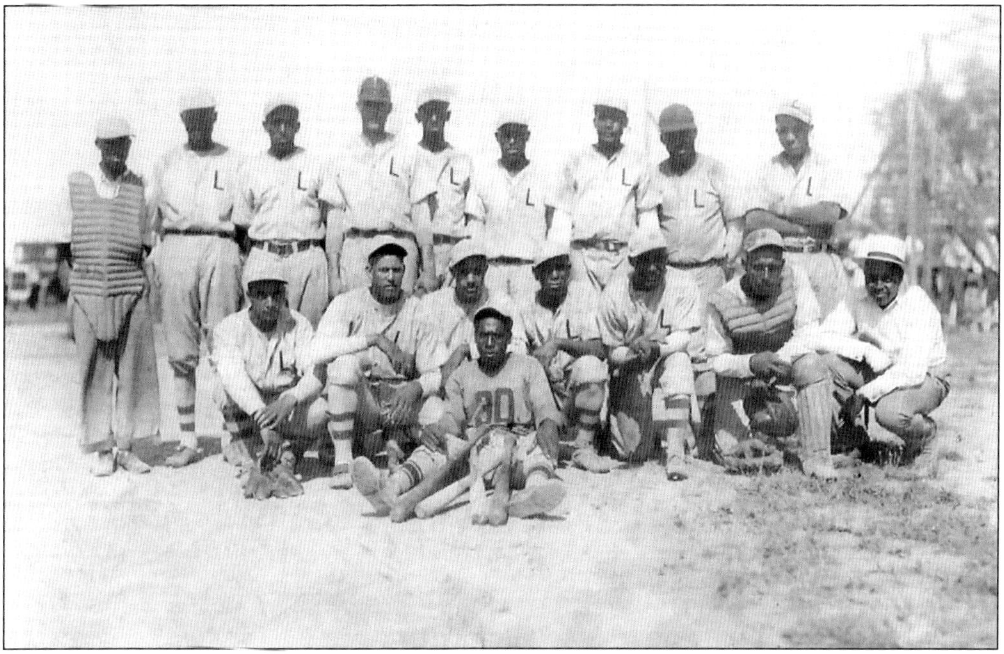

The Lawrenceburg Athletics was an African American baseball team featuring local and touring professionals. The team poses here before a game around 1930. (Courtesy of Birdwhistell collection.)

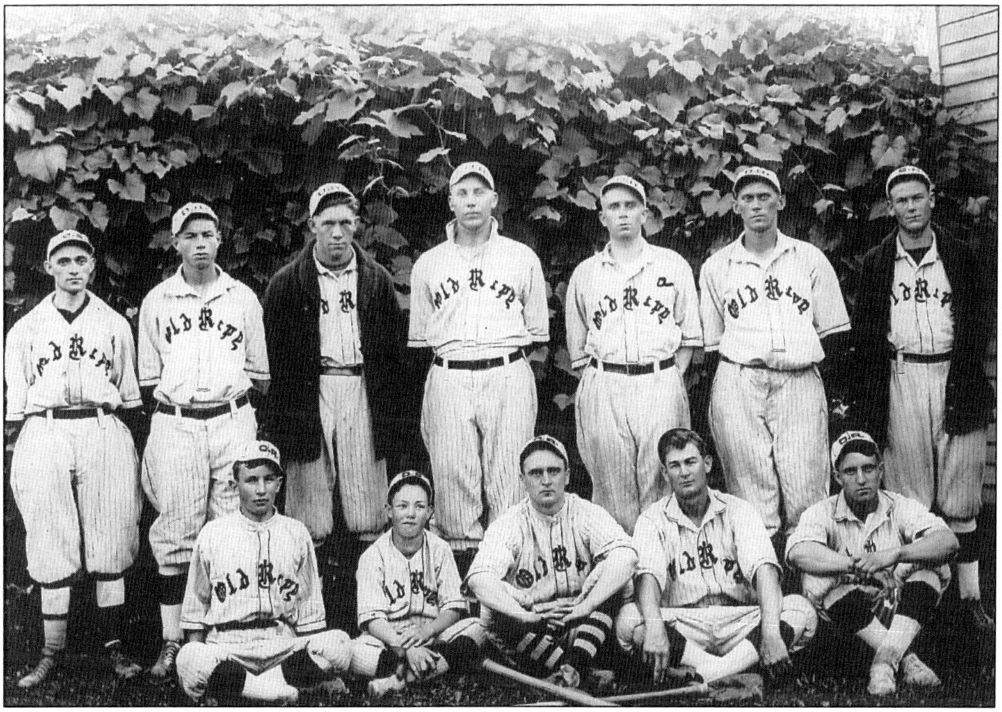

In 1915, the Ripy baseball team of Tyrone played a challenging schedule. The players are, from left to right, (first row) Eula Bercaw, C.T. Ward, J. McAleese, Howard Hampton, and Allison "Toots" Nowlin; (second row) Bill Proctor, Edward Howard, Fred Purcell, Tom Watson, Garry Bottoms, Tom Chapman, and Zip Hampton. (Courtesy of Birdwhistell collection.)

PROGRAM

Lawrenceburg

Baseball Club

Lawrenceburg, Kentucky

1951 Season

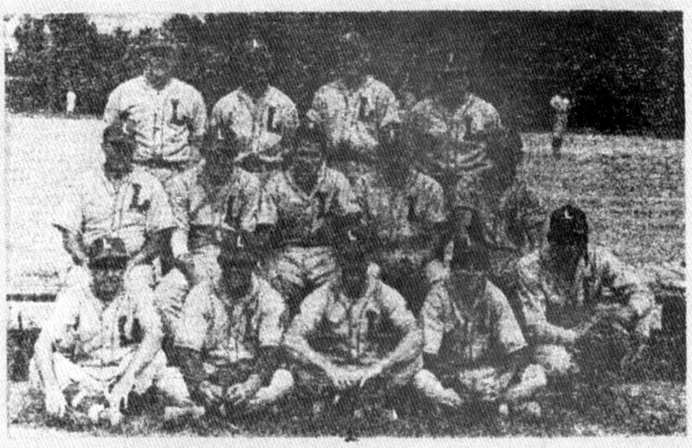

Lawrenceburg Team—1949

Lawrenceburg's baseball club was in the Bluegrass League and had a large hometown following, with as many as 400 people attending games. In the 1951 season, the club used sponsors to print a program for each home game. The programs included a short biographical sketch of a selected player each week. (Courtesy of Bryant collection.)

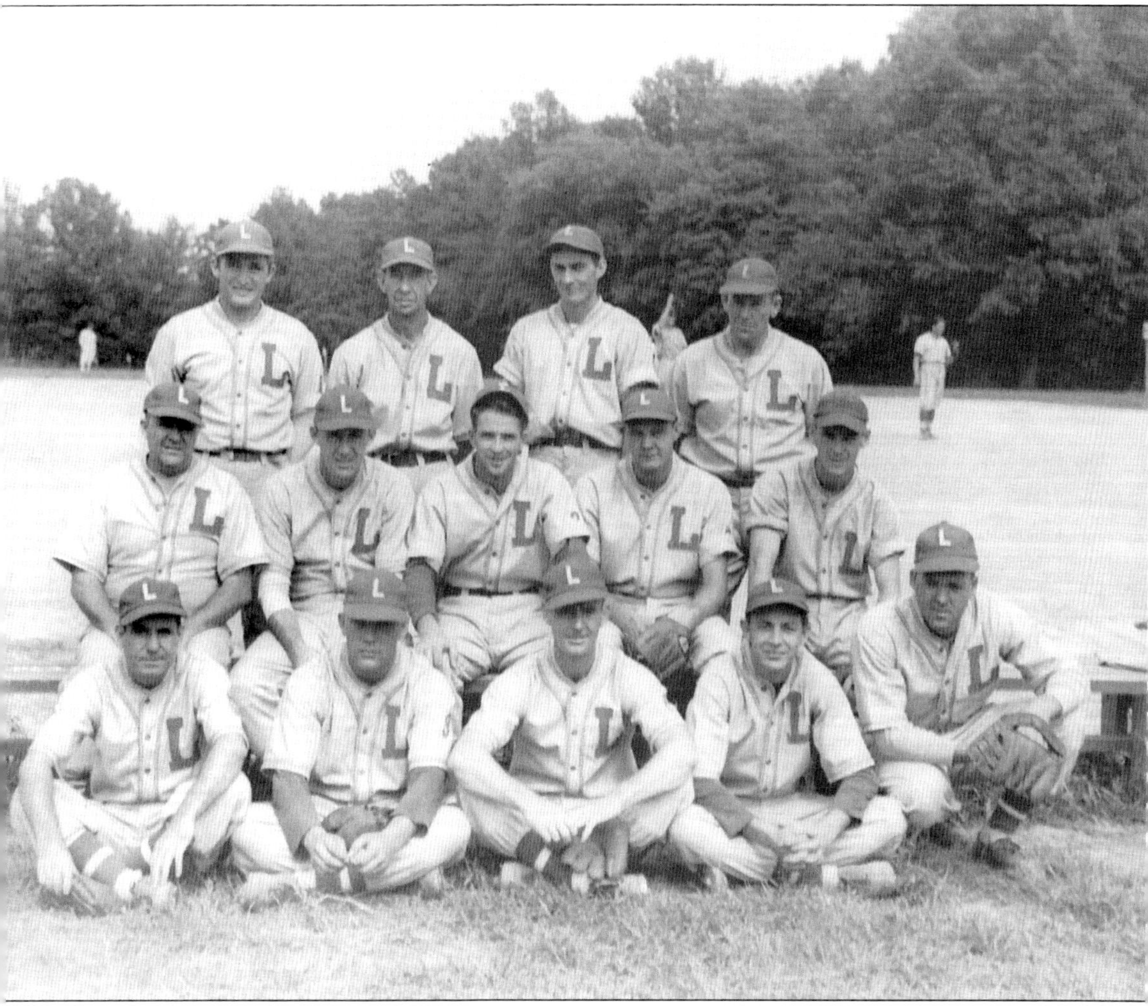

The Lawrenceburg Laries baseball team was started in 1947 and included many World War II veterans. The 1949 team included, from left to right, (first row) Meb Hanks, Willard Bryant, J.T. Hanks, Gilbert Bryant, and ? Taylor; (second row) Bill Baker, Floyd Moffett, Zeke Inman, Alvin Wicke, and unidentified; (third row) Jim Simpson, Denver Qualls, Mide Green, and Keith Allen. (Courtesy of Bryant collection.)

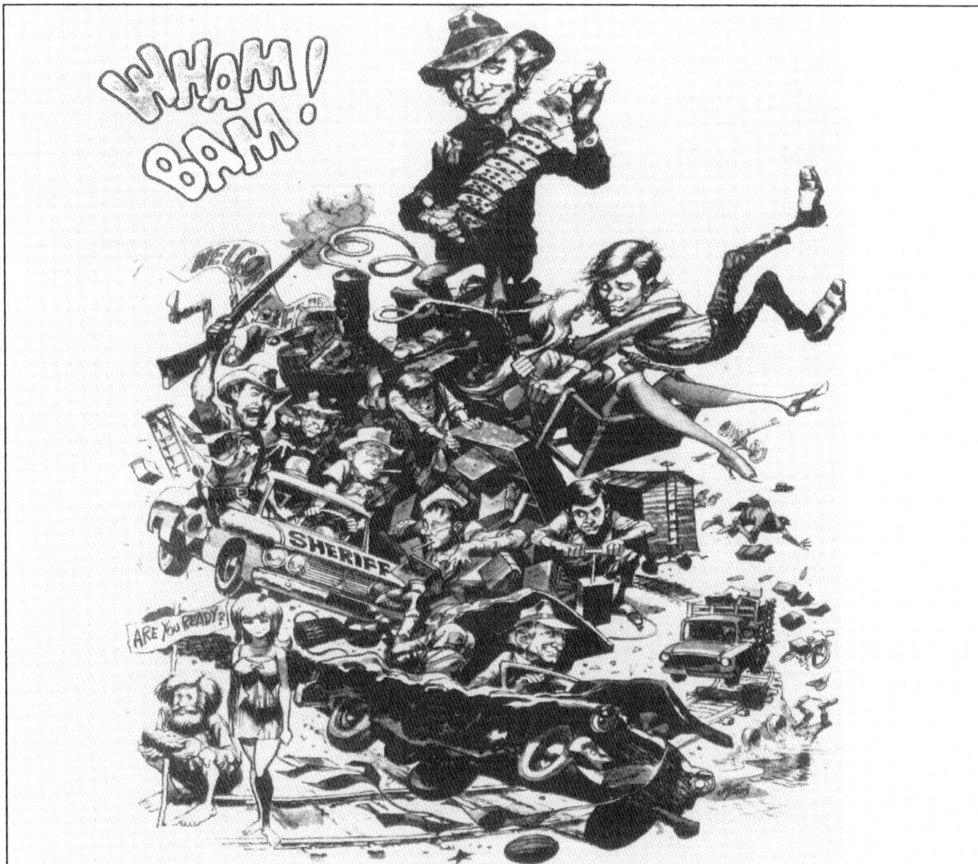

In 1966, Hollywood came to Lawrenceburg to make the movie *The Flim-Flam Man*. The movie starred Academy Award winner George C. Scott and a number of other notable character actors. The production also involved several local Lawrenceburg residents in background roles. (Courtesy of Anderson County Historical Society.)

Eight

Social and Religious Life

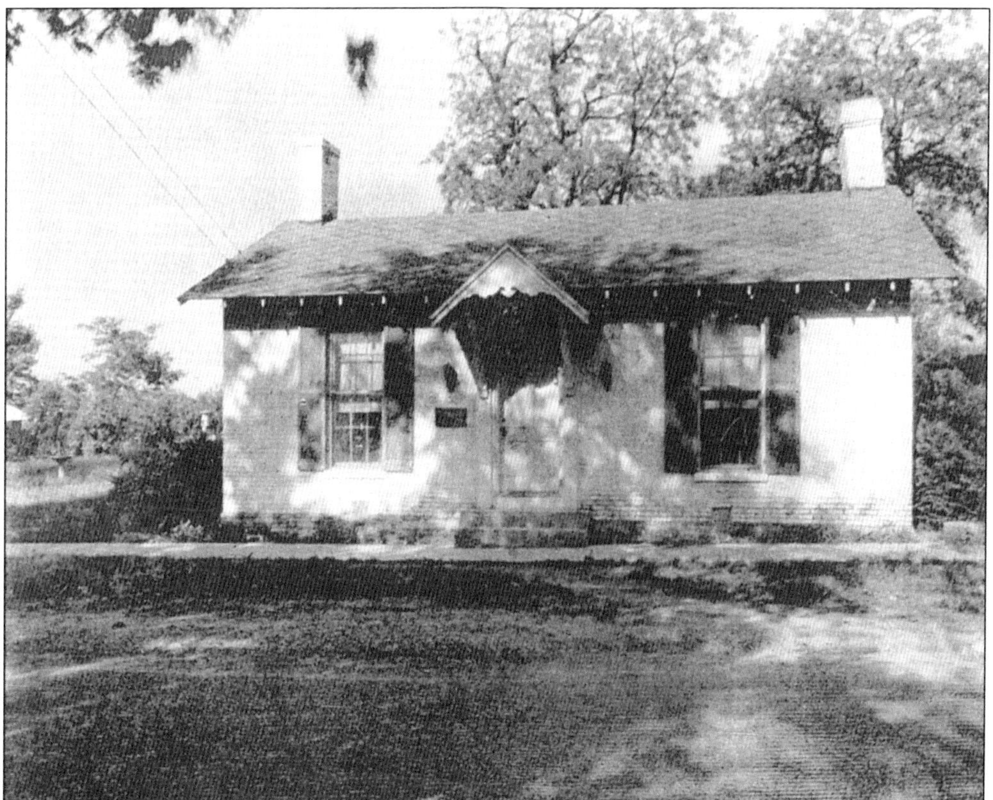

The Pierian Clubhouse, the oldest house in Lawrenceburg, was built in 1818. It was the home of the Lawrenceburg Pierian Woman's Club. Today, it stands on its original site on North Main Street but has been incorporated into the Anderson County Library building. (Courtesy of Anderson Historical Society.)

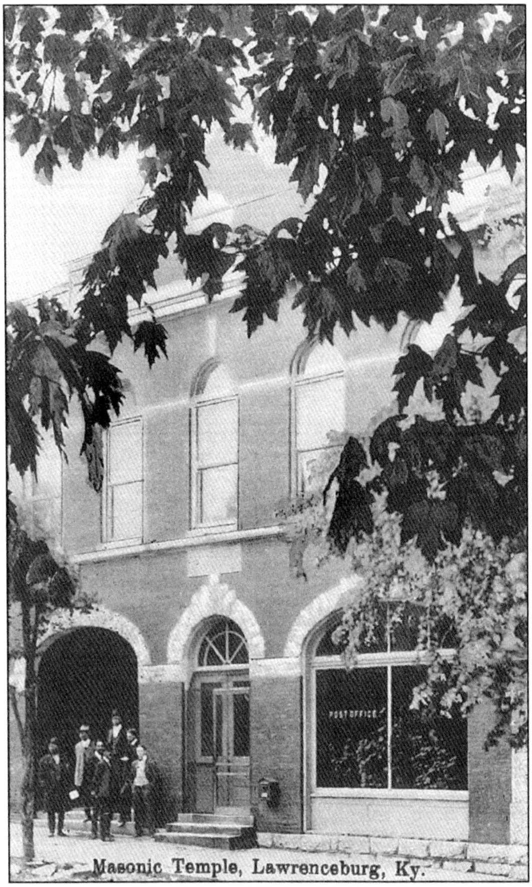

The Masonic Lodge was granted a charter in 1828. In 1906, a new building was constructed to house Masons, the Eastern Star, and the Agricultural Extension offices. At one time, it also served as the post office. (Courtesy of Hawkins collection.)

In 1900, three of the churches in downtown Lawrenceburg looked different than they do today. Fires and windstorms damaged the buildings, and they were rebuilt in new styles. One, the First Christian Church, is at a different location as well. (Courtesy of Hawkins collection.)

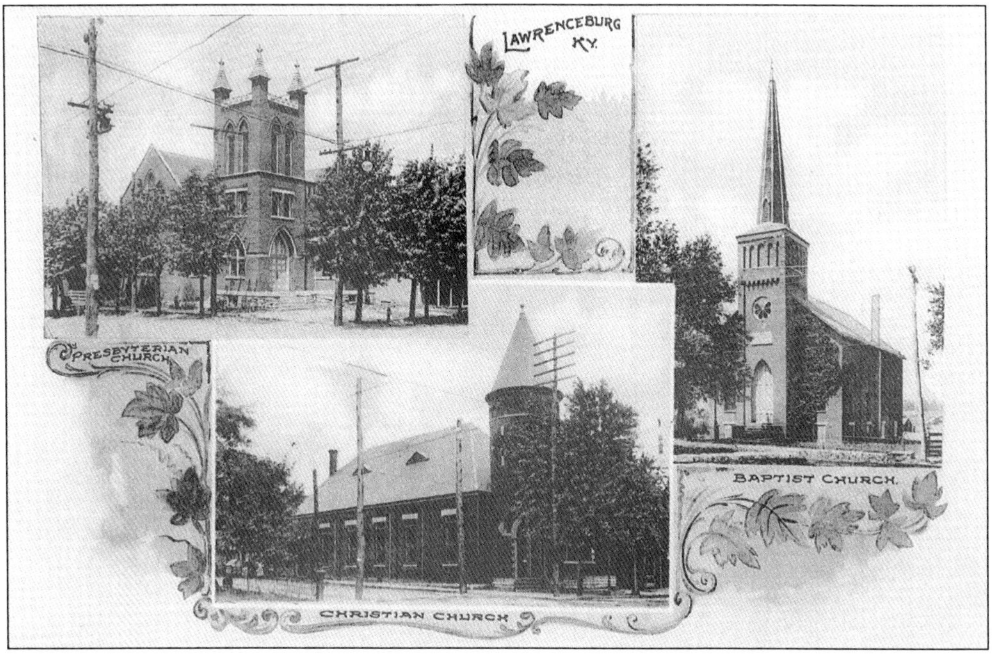

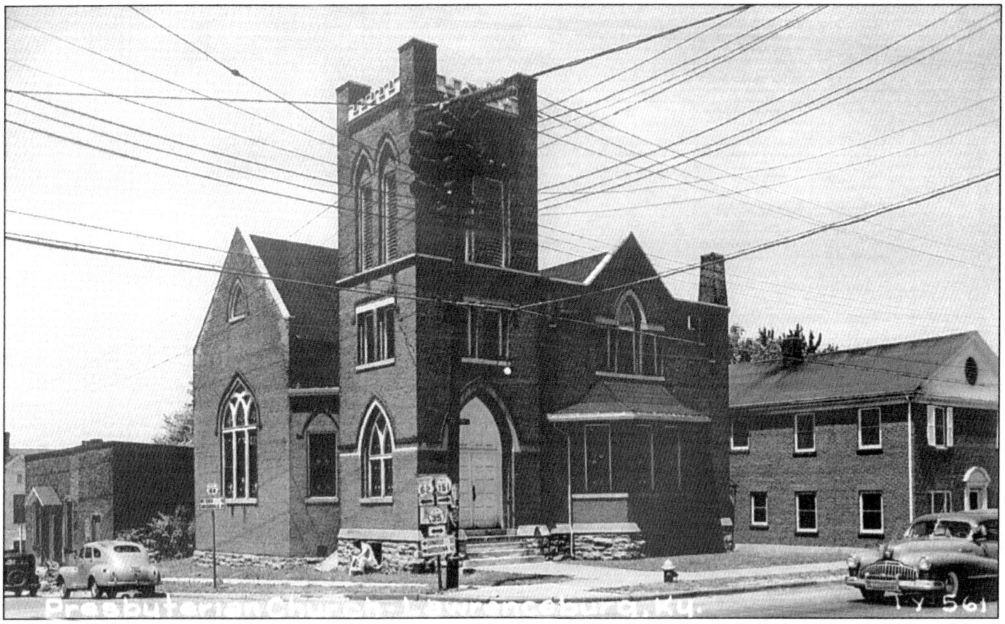

The First Presbyterian Church was organized in 1828, and a small brick building was constructed in 1834. In 1858, windstorms destroyed the building on Woodford Street, and it was rebuilt but later hit by a second windstorm. In 1905, the church caught fire and was completely destroyed. The present church building was erected in 1906. (Courtesy of Hawkins collection.)

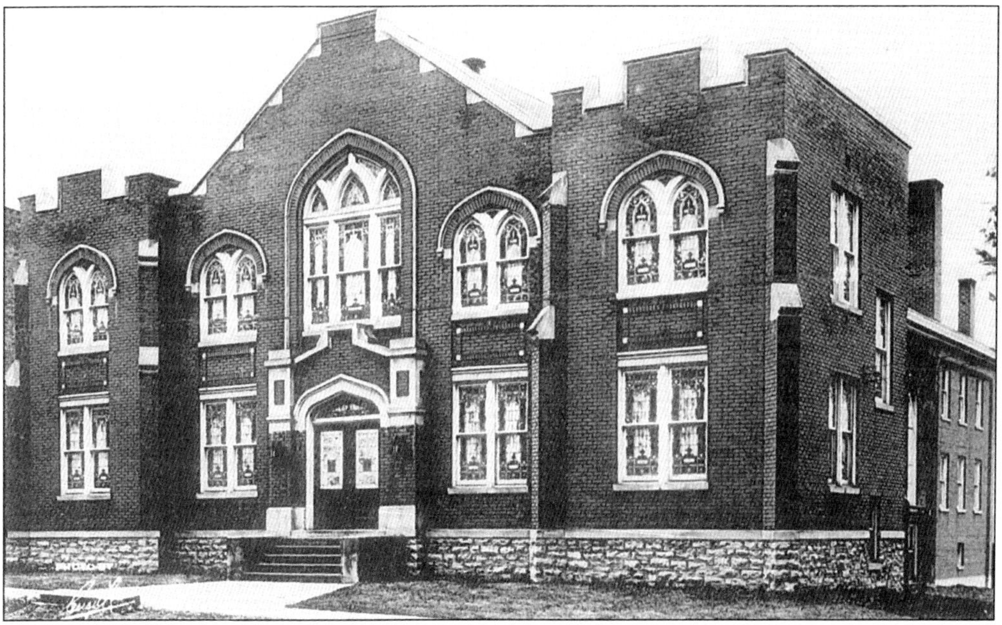

The First Baptist Church was begun in 1834. In 1870, a new church was completed on the land where a new church now stands. On January 8, 1924, fire severely damaged the original church but the walls remained. The remodeled building lacked the tall steeple and bells. (Courtesy of Hawkins collection.)

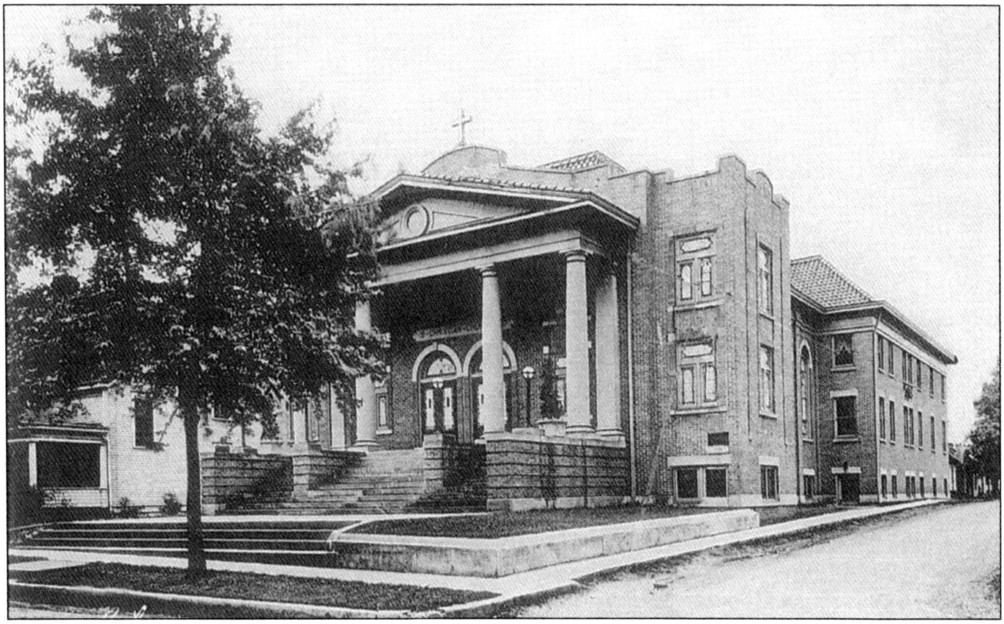

The First Christian Church was destroyed by fire in June 1876. In 1879, a new brick church replaced the original. That building, at the corner of Main and Court Streets, then burned in 1899, and a new church building on the same site burned again after it had been sold. The present church was dedicated on August 28, 1921. (Courtesy of Hawkins collection.)

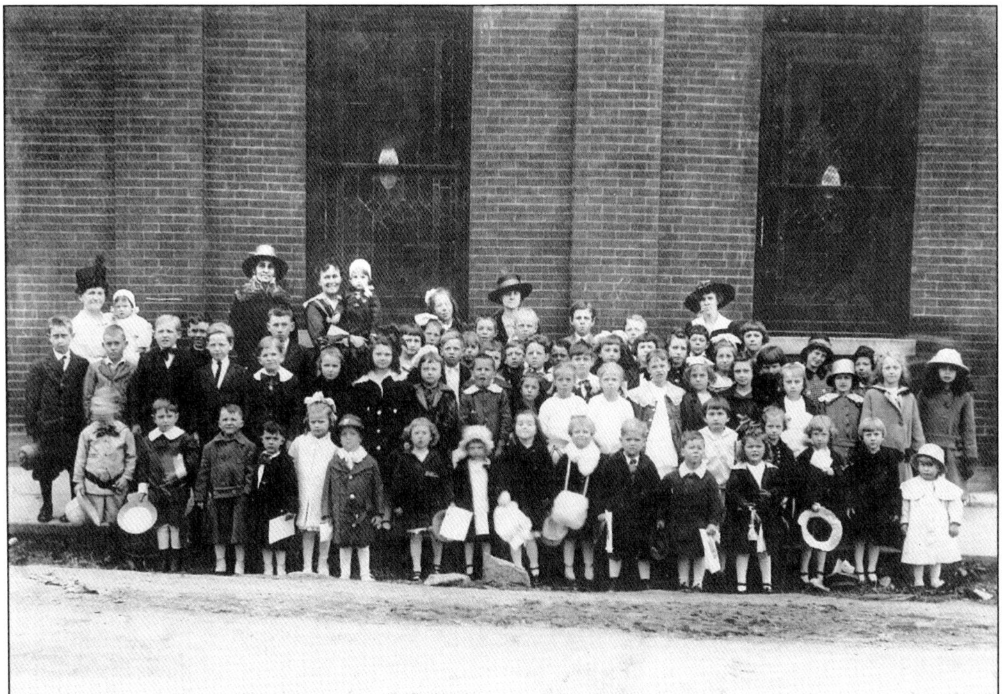

Sunday school classes pose for a photograph outside the First Christian Church, which at the time was on the corner of Main and Court Streets. Today, the Feland Building occupies the site. (Courtesy of Hawkins collection.)

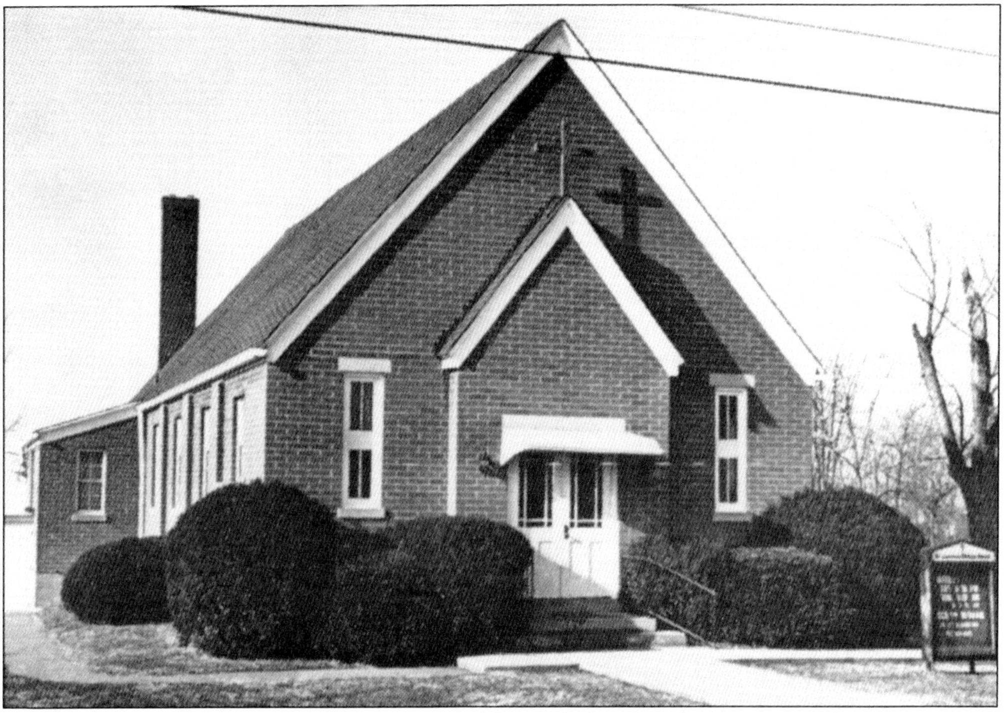

In April 1873, land was purchased for a possible Catholic church in Lawrenceburg. In 1894, a building on Gatewood Street was purchased and converted into a church, known as Saint James of Less. That building was torn down in 1939, and homes were used for church services. Mass was said on the second floor of Klink's drugstore for a time. The current church building was completed in 1952. (Courtesy of Litkenhus collection.)

The Lawrenceburg Methodist Church was located on South Main Street across from the First Christian Church. It was first a wooden structure, but the outside was later bricked. Due to a generous donation, the old church was sold and a new building was constructed and dedicated at the corner of Carlton Lane and the Highway 127 bypass. (Courtesy of Ramona Simpson.)

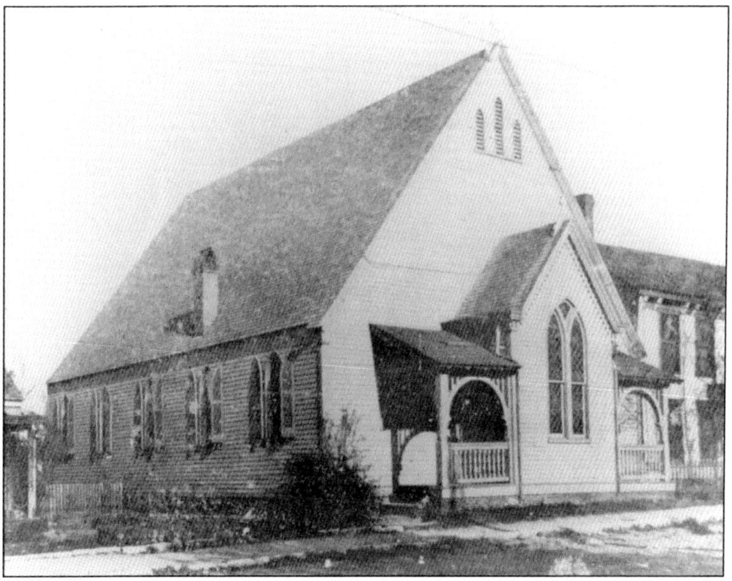

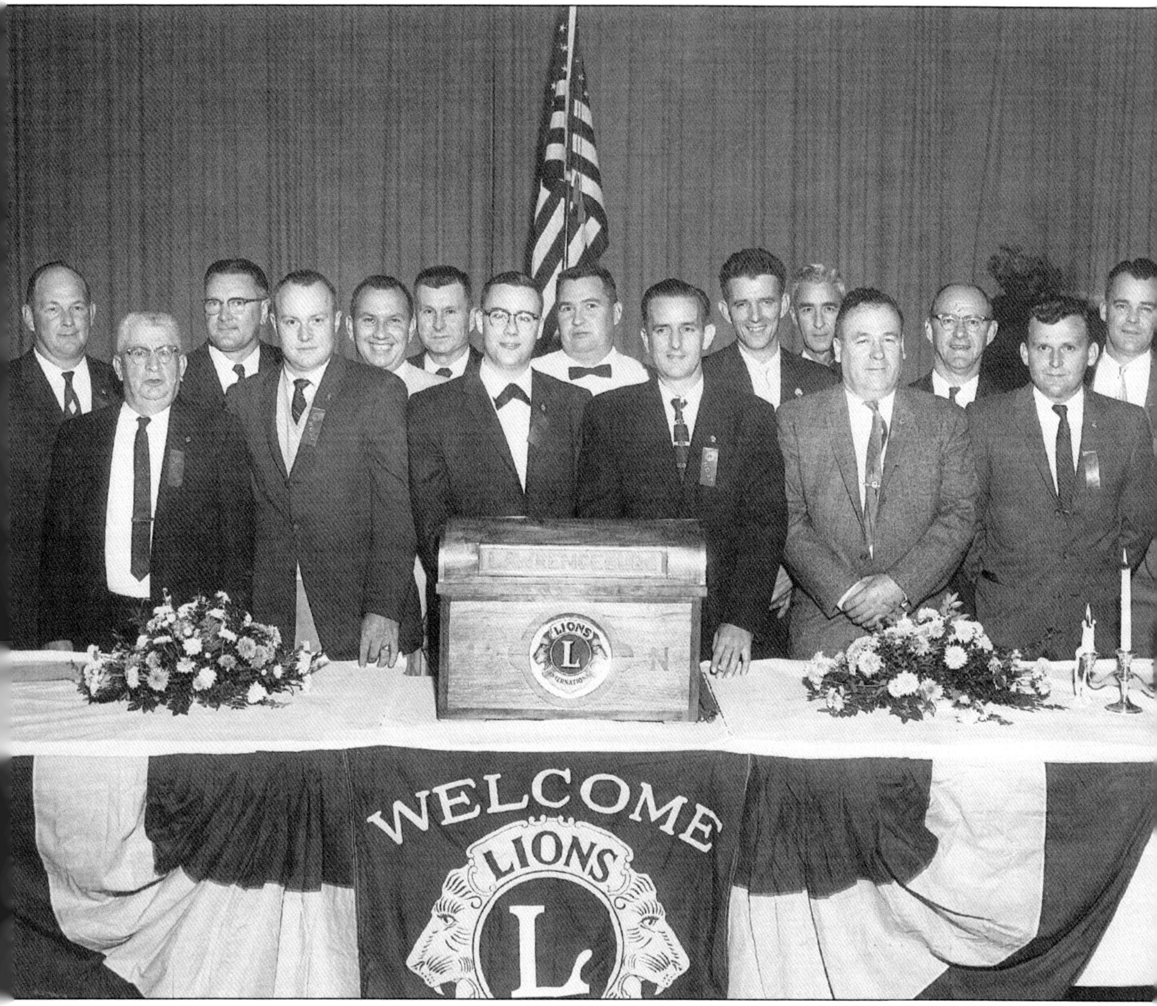

The men of Lawrenceburg have always been civic-minded and associated with organizations that support worthy causes. In Lawrenceburg, those organizations have included the Rotary Club, Lions Club International, and the Jaycees. The Lions Club has supported eye research, especially for children, among other things. Members of the local chapter are seen here in the late 1950s. (Courtesy of Phillips collection.)

Nine

EDUCATION

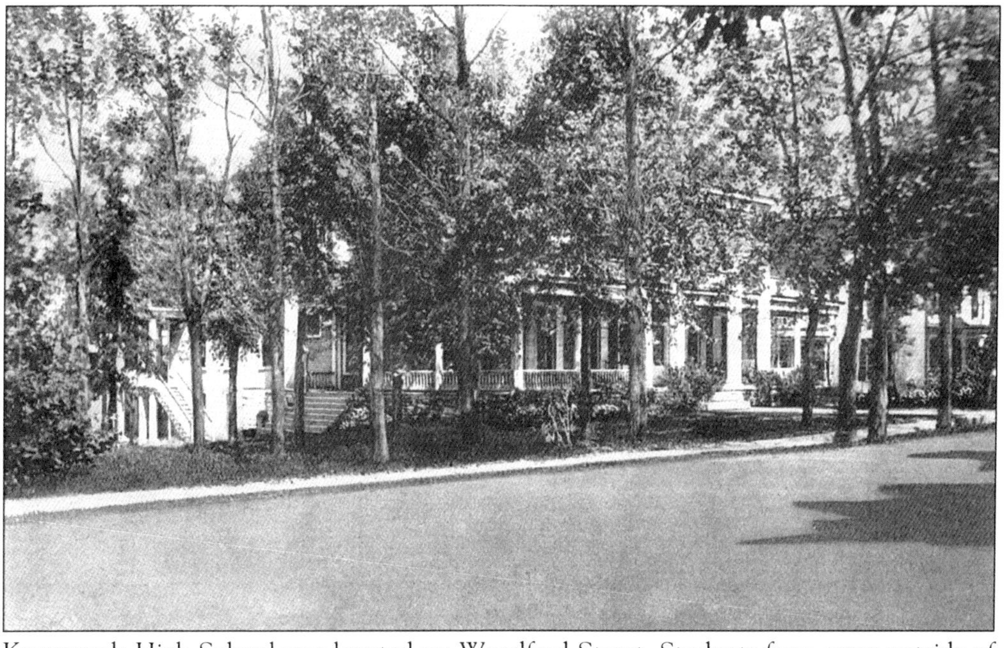

Kavanaugh High School was located on Woodford Street. Students from areas outside of Lawrenceburg often attended Kavanaugh while boarding with residents of Lawrenceburg. The Kavanaugh students were quite successful and were accepted into military academies; to many, the school was known as "Little Annapolis." The Kavanaugh School is in the National Register of Historic Places, and a historical plaque on the grounds lists its most famous alumni who distinguished themselves with military and often naval careers. (Courtesy of McAnly collection.)

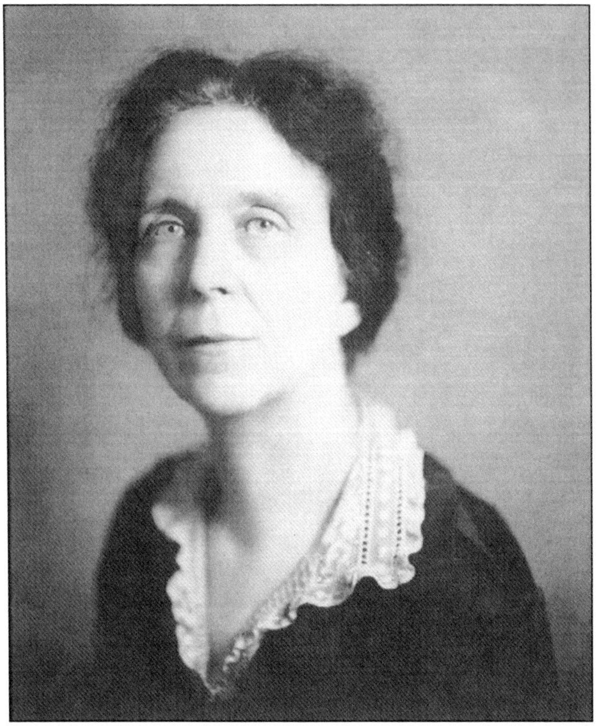

Rhoda Kavanaugh was an inspirational teacher and educator. She was a graduate of the University of Kentucky and provided strong leadership to other teachers as well as students at the high school. She was married to Dr. Charles Kavanaugh, and she developed the Kavanaugh Academy and High School in their home. Mrs. Kavanaugh was also known for her participation at high school basketball games, where she challenged players to do their best. (Courtesy of Anderson County Historical Society.)

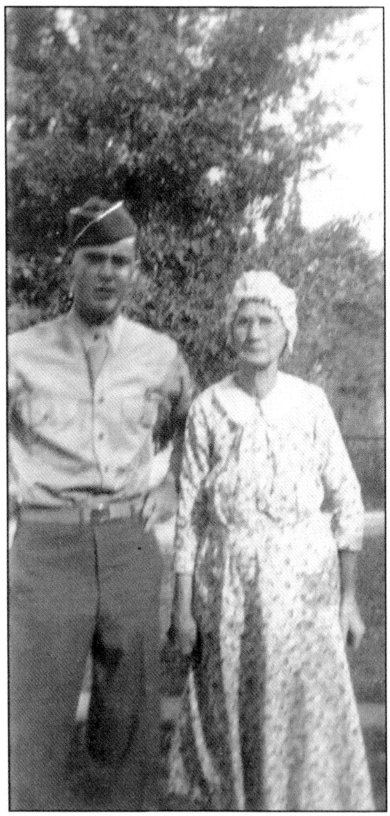

Wilton A. Hawkins is shown with his grandmother Mertie around 1945 while on leave during World War II. After being discharged from the Army, Hawkins transferred from the University of Kentucky to Carnegie Tech (now Carnegie Mellon University) in Pittsburgh to complete his engineering degree. After graduating in 1948, he joined the plastics department at DuPont and concentrated on research and experimentation with Teflon. By 1953, Hawkins had started his own company, Chemplast, Inc. and developed and patented a wide variety of products and processes. Chemplast expanded from the 1950s through the 1970s, and products ranged from those used in aerospace to medicine. Hawkins maintained involvement in his companies as president or vice president of certain product lines. He and his wife, Teddy, became involved in many philanthropic endeavors, but he never forgot his Kentucky roots and generously supported the Anderson County Historical Society and Museum, Anderson County Board of Education, Anderson County Library, Anderson County Scholarship Fund, and Lawrenceburg Methodist Church, among others. Hawkins has received many honors in recognition of his professional and philanthropic efforts. In 1997, Carnegie Mellon University awarded him with an honorary doctorate. (Courtesy of Hawkins collection.)

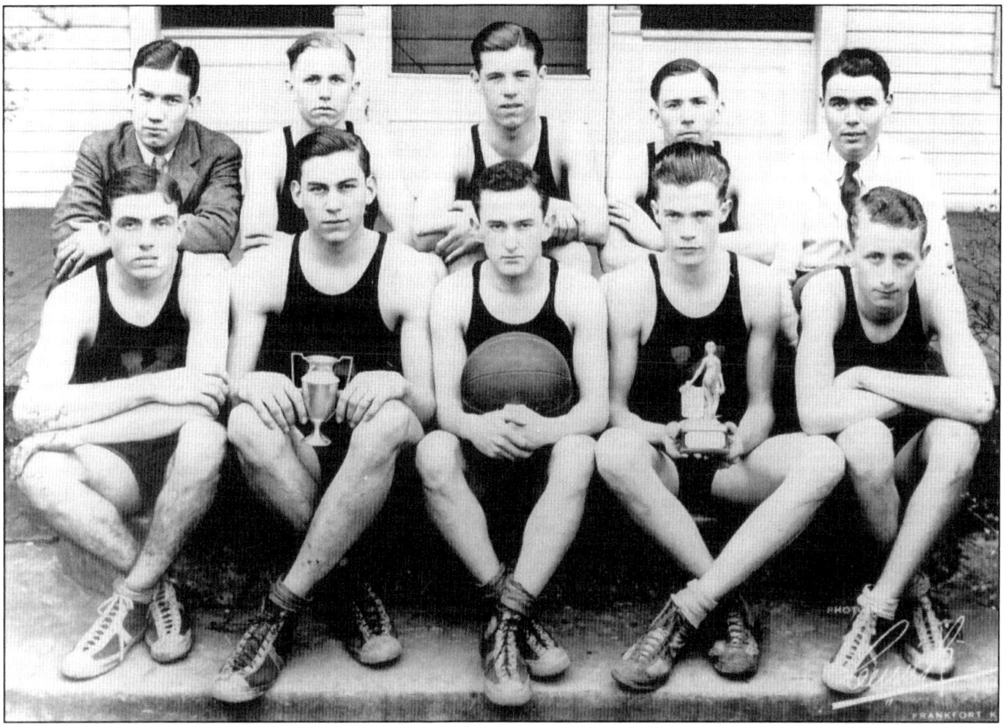

Forest "Aggie" Sale (first row, second from the left) was a standout basketball player at Kavanaugh. He was six feet, four inches tall and used his height to his advantage. In college, he was an All-American at the University of Kentucky. Earle Jones (in suit) was a successful coach at Kavanaugh and, after leaving there, led Maysville High School to the Kentucky state championship. (Courtesy of Birdwhistell collection.)

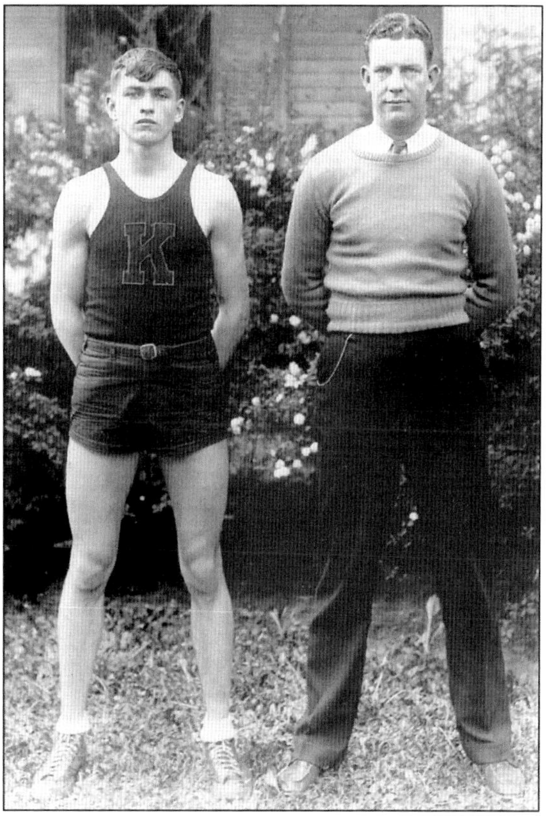

Ralph Carlisle (left), an all-SEC player for the University of Kentucky in 1936 and 1937, and Paul McBrayer, who had been an All-American in his younger days and became the coach at Kavanaugh, pose together in this 1932 photograph. Both men coached at Kavanaugh, and Carlisle was considered one of Kentucky's best coaches, winning three state championships at Lexington-Lafayette. McBrayer was an assistant at the University of Kentucky under Adolph Rupp and went on to become the winningest coach in Eastern Kentucky University's history. (Courtesy of Birdwhistell collection.)

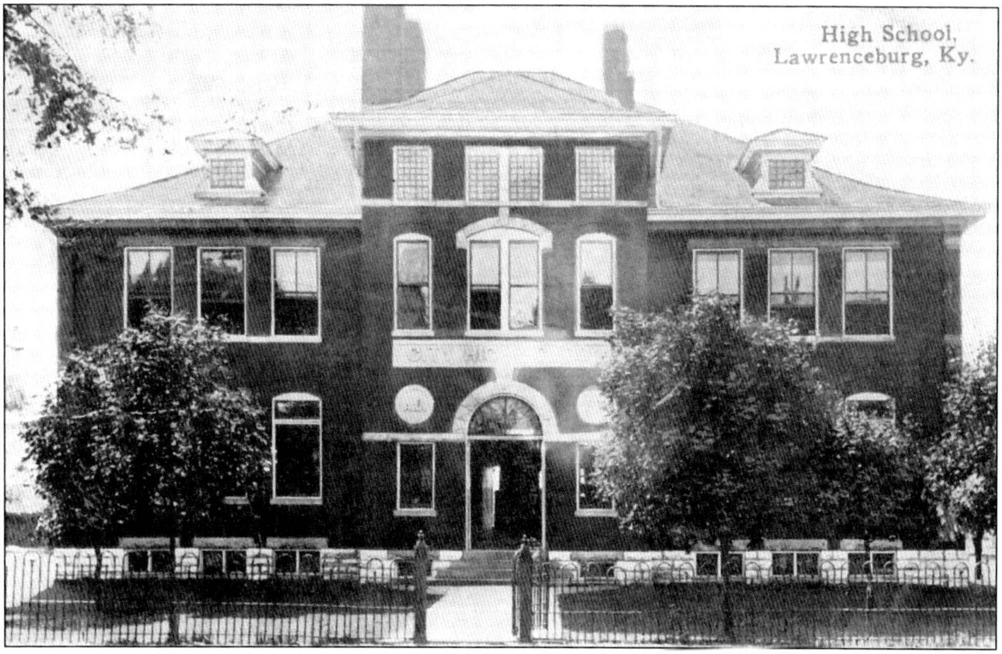

Lawrenceburg High School was built in 1904 and served the city as both an elementary and secondary school. It was located at the corner of Woodford and College Streets. After the high school relocated to the old Witherspoon mansion on North Main Street, only elementary students remained. Kavanaugh began as a private academy, expanded to become a private high school, and eventually became the public county high school. Both schools maintained high academic standards and were successful in athletics, especially basketball. (Courtesy of Bryant collection.)

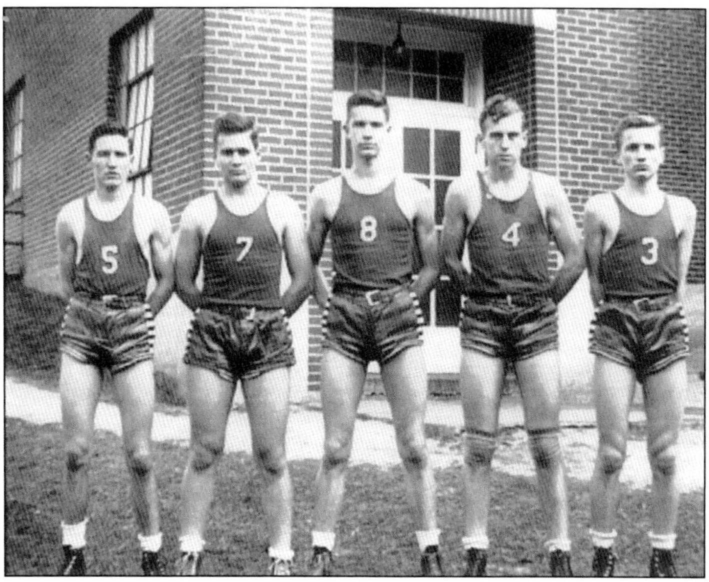

The starters on the 1943–1944 Lawrenceburg High School basketball team seen here are, from left to right, Bobby Hyatt, Charles Humston, Bobby Paasch, Billy Spencer, and Eddie Humston. Humston was an all-state player who went on to a career in medicine. (Courtesy of Birdwhistell collection.)

Jimmy Dan Conner was a three-time all-state basketball player at Anderson High School. In 1971, he was Kentucky's Mr. Basketball and a *Parade* and scholastic All-American. He went to the University of Kentucky, where he was a three-year starter and a member of the 1,000-point club. (Courtesy of Frances Conner Gorham.)

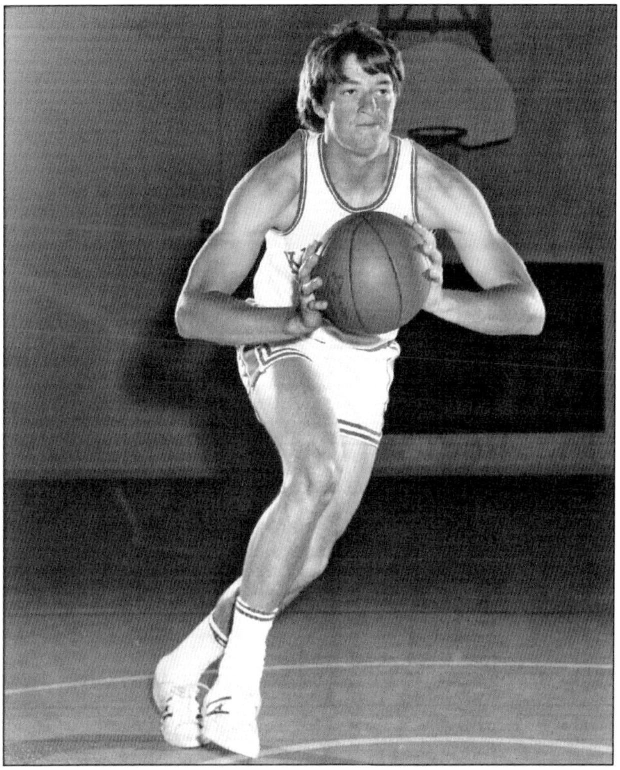

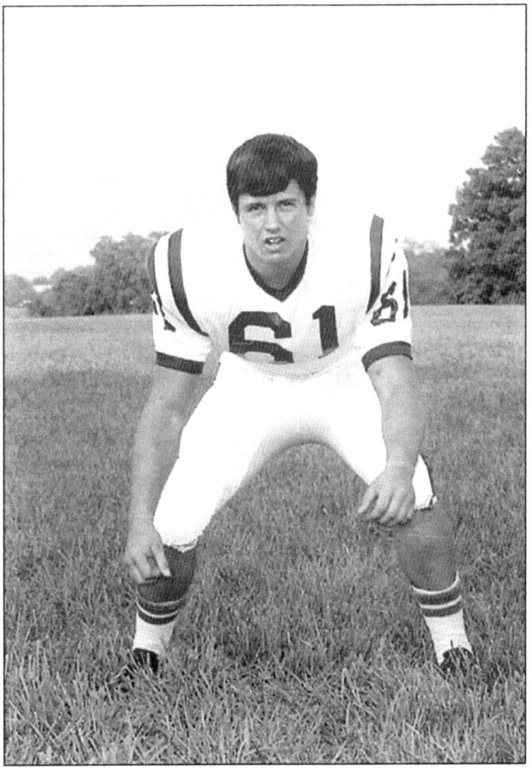

Bobby Ware was a two-time all-state football player in the late 1960s. He received a scholarship to play at the University of Kentucky, but because of his diabetes he was unable to play. He earned his bachelor's and master's degrees in civil engineering at Kentucky. (Courtesy of Pam Ware.)

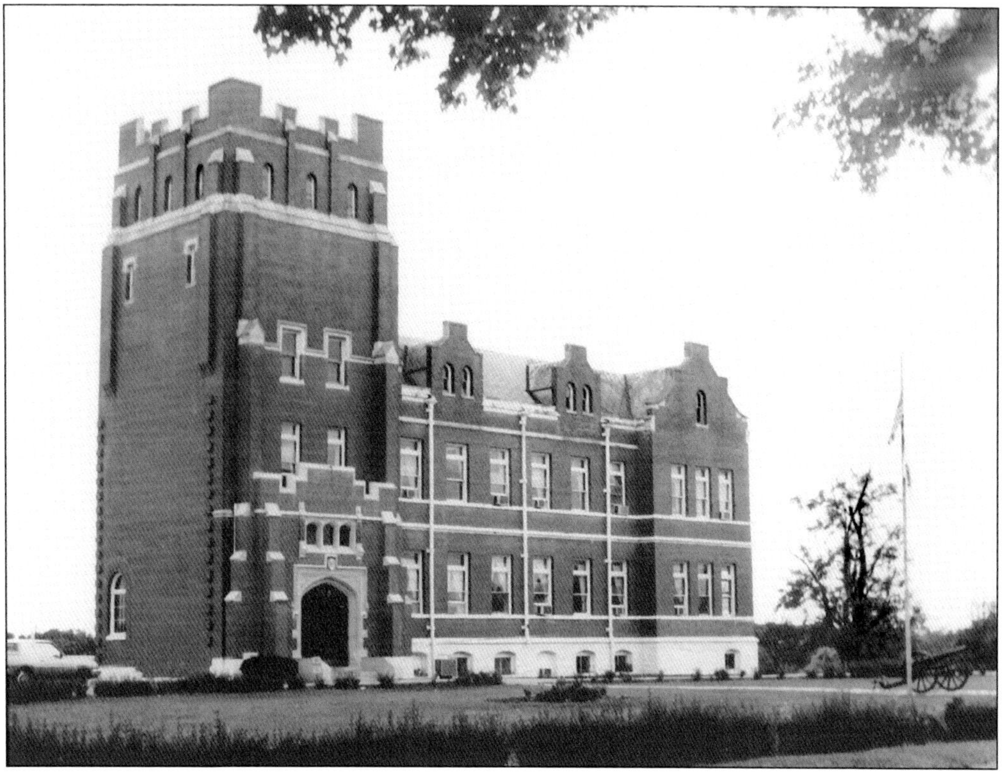

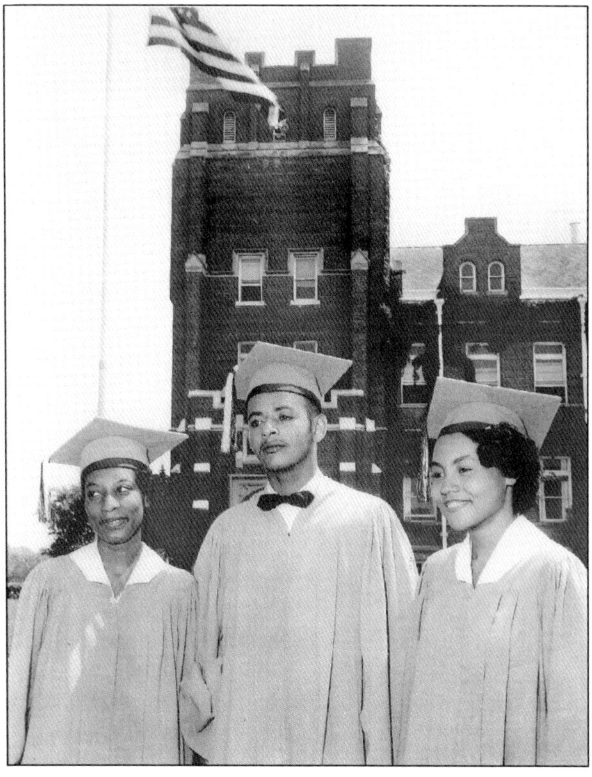

In 1902, the Day Law went into effect, segregating Kentucky's schools. After the ruling, Berea College, an integrated college, supervised the development of Lincoln Institute, an all-black institution in Shelby County. The main administrative building at Lincoln Institute was named Berea Hall. (Courtesy of Gary Brown.)

Of the three honor students seen here at Lincoln Institute, two were from Lawrenceburg: salutatorian John Brown and, on the right, valedictorian Marva Cunningham. In 1962, segregation was declared unconstitutional, and the schools in Anderson County were integrated. (Courtesy of Gary Brown.)

William H. Townsend was a native of Glensboro who received his law degree from the University of Kentucky and became a successful corporate attorney. He was also an Abraham Lincoln scholar, writing a number of books about the former president's life. (Courtesy of Anderson County Historical Society.)

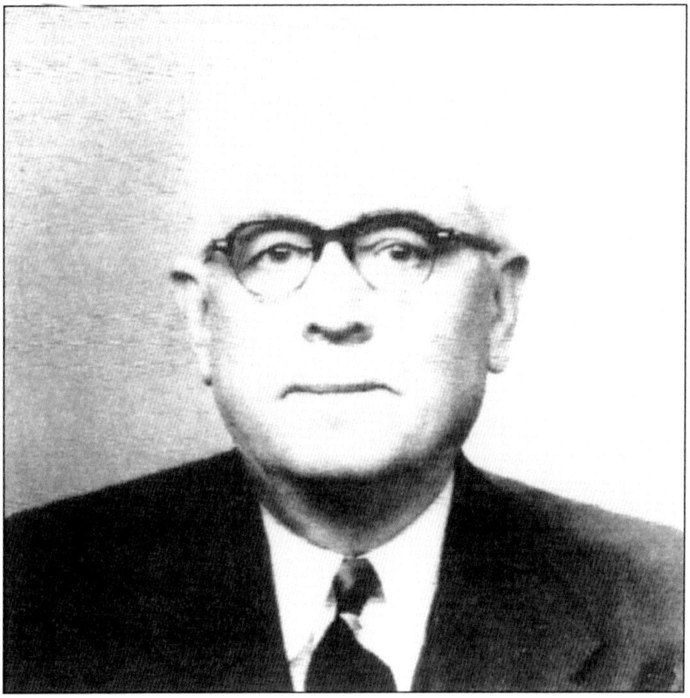

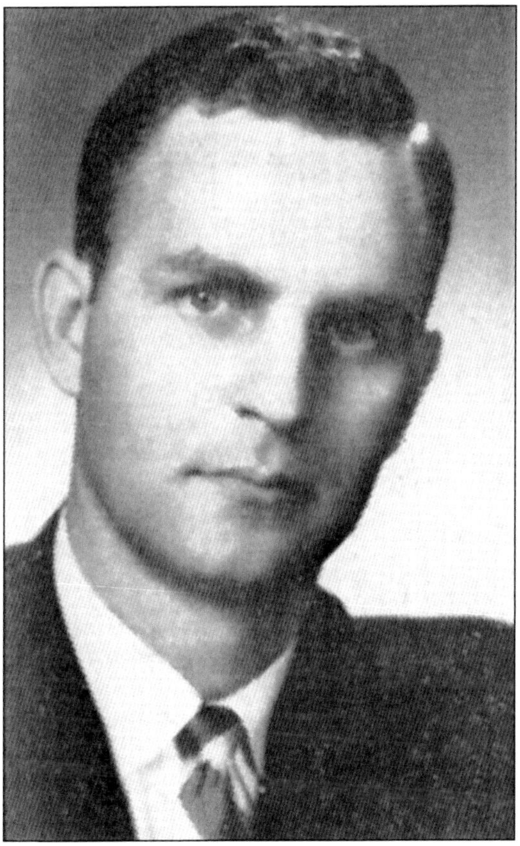

Walter Patrick is listed here not only because of his academic credentials in law but as the chairman of the Anderson County Scholarship Fund. Only the interest from this fund is used for scholarships to students from Anderson County. (Courtesy of Hawkins collection.)

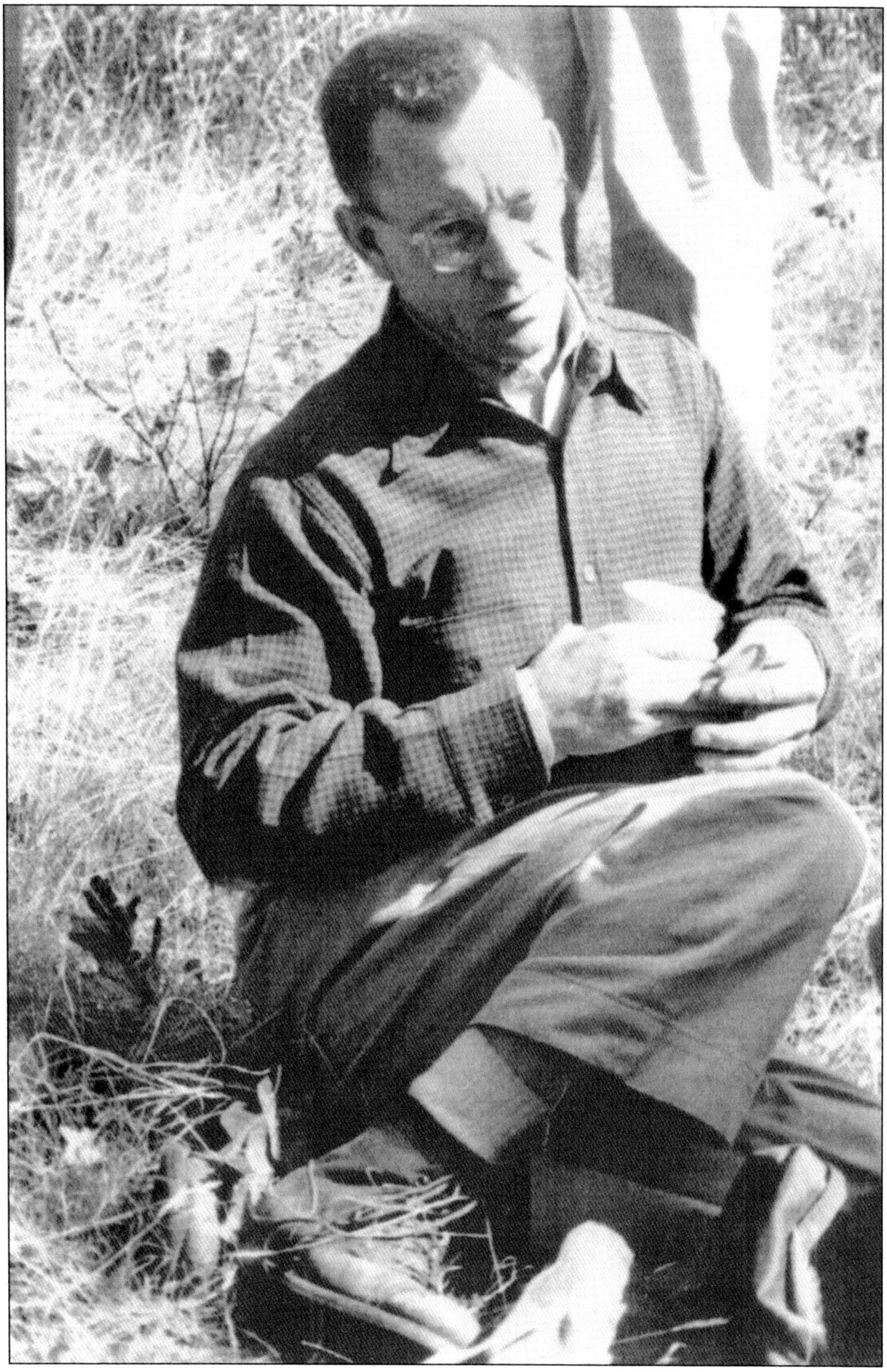

John Campbell Goodlett is a good example of a man of the "Greatest Generation." He graduated from Lawrenceburg High School and attended the University of Kentucky, where he was drafted by the Army in World War II. After the war, he returned to college, graduated Phi Beta Kappa, and attended graduate school at Harvard University, where he earned a PhD. He taught in the geography department at Johns Hopkins University but died of a heart attack one month short of his 45th birthday. (Courtesy of Bryant collection.)

Ten

LEADERSHIP

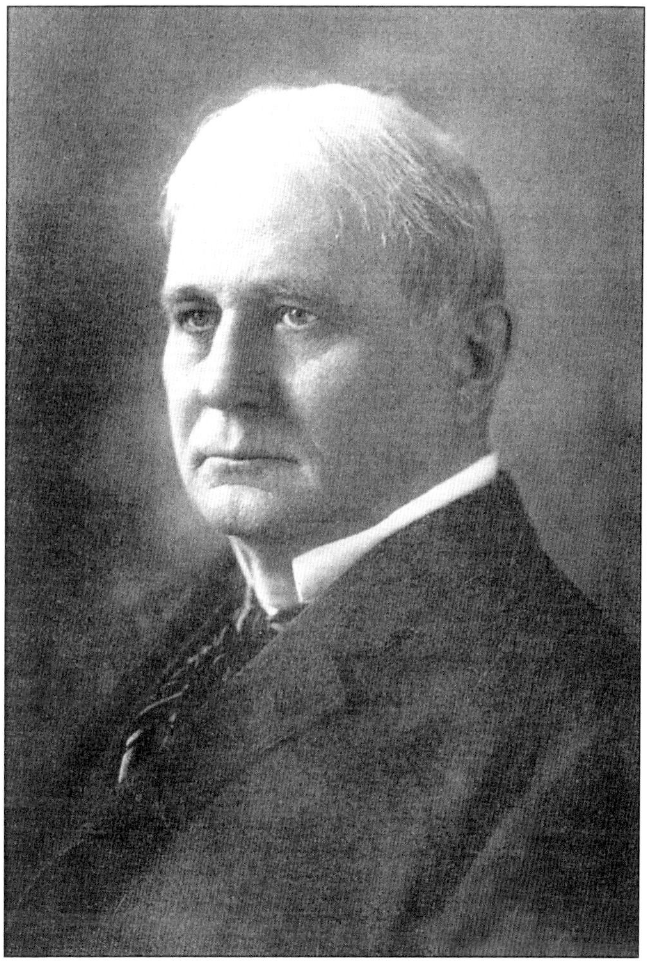

Early in the 1900s, business and other community leaders were active in promoting Lawrenceburg's development. They worked for the expansion of essential services and the improvement of safety and recreational activities for local citizens. The American Legion and public-spirited doctors, lawyers, and teachers assumed leadership roles to move the town forward. Champ Clark, christened James Beauchamp Clark and born in Anderson County on March 7, 1850, began his career as a teacher in Anderson County. After graduating from the University of Cincinnati Law School, he started on the road to politics. After relocating to Missouri, Clark ran for Congress and became a member of the US House of Representatives. By 1912, he was speaker of the house and later a presidential candidate. (Courtesy of Bryant collection.)

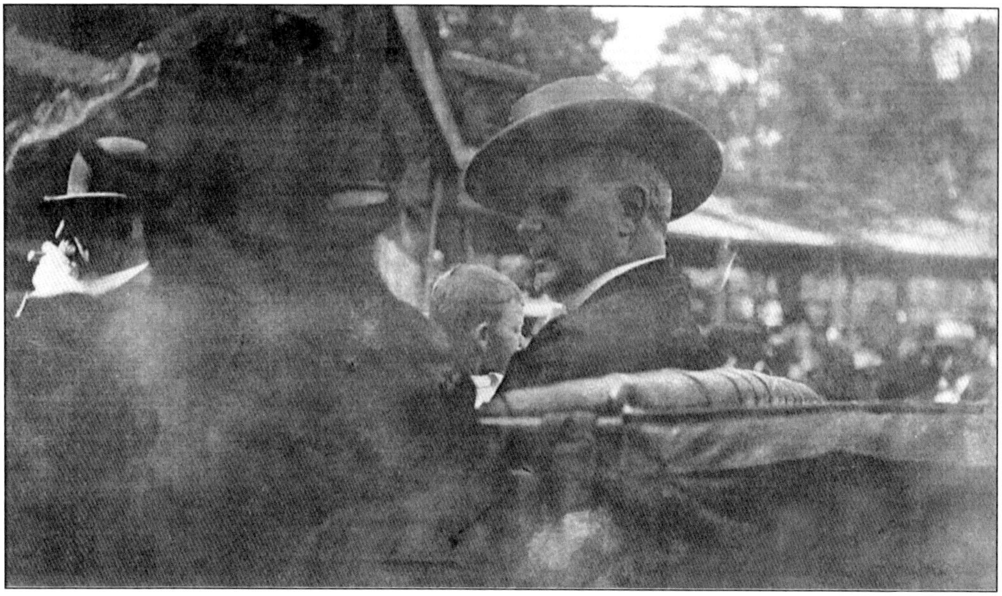

In 1910, Champ Clark, on a speaking tour of Kentucky, came back to Lawrenceburg for a political rally at the old fairgrounds. The governor, James B. McCreary, was also in attendance. A group of representative citizens met the distinguished visitor at the train. Clark visited his mother's grave and met with locals during his visit. (Courtesy of Wilhoit collection.)

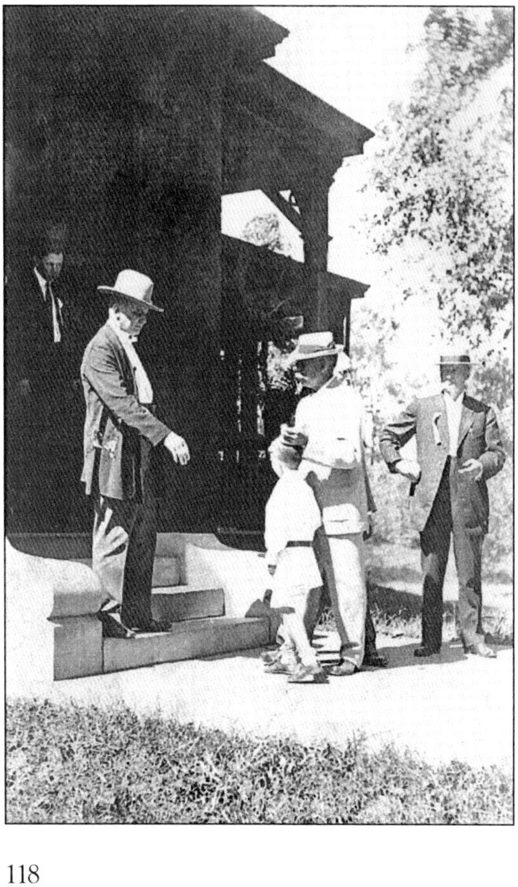

Champ Clark (left) was escorted to the Jesse M. Johnson home on South Main Street during his 1910 visit. Here, he stands in front of the Johnson house with Johnson (in the white suit). (Courtesy of Wilhoit collection.)

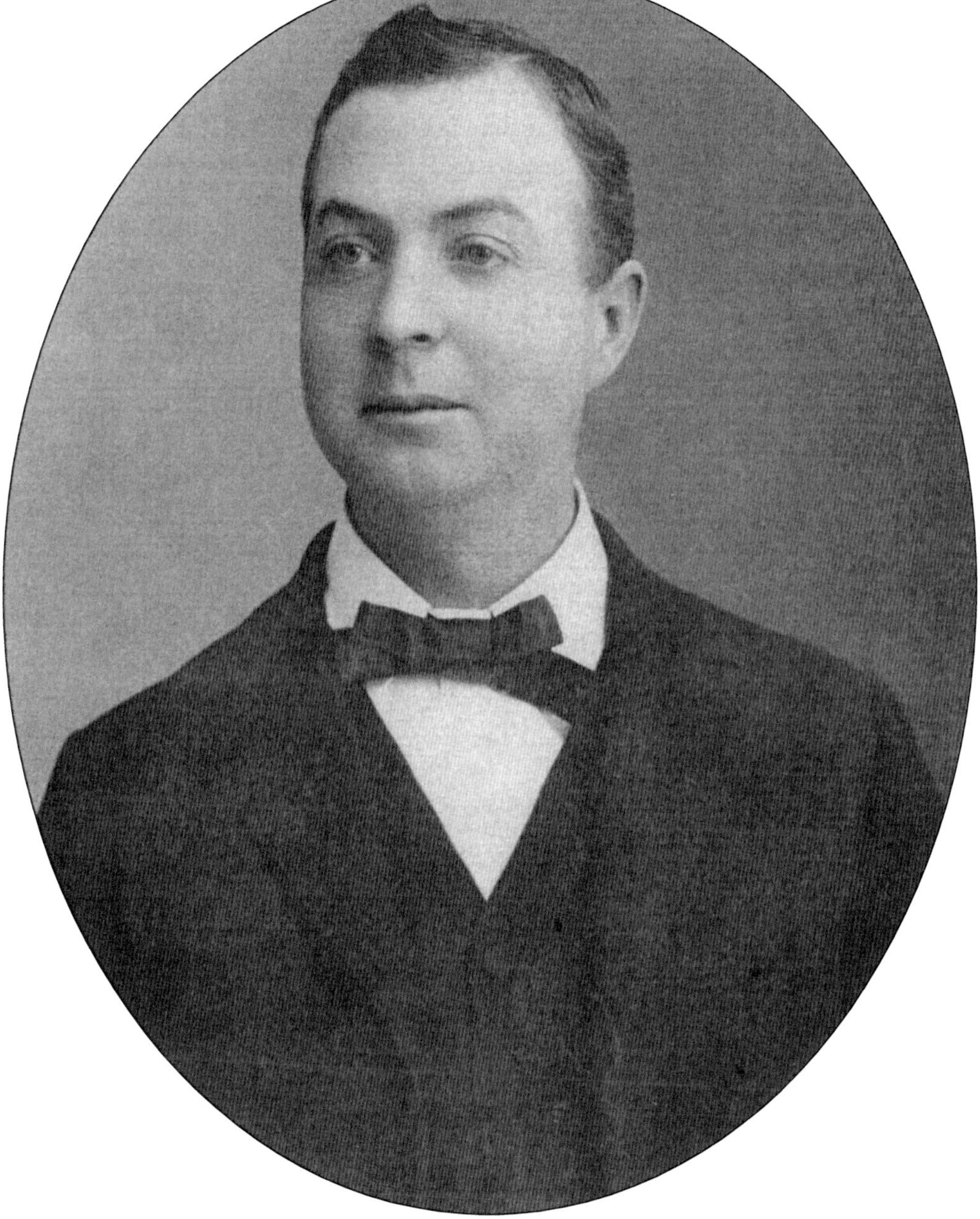

Thomas J. Ballard was born in Shelby County and moved to Lawrenceburg as a young man. Once here, he became involved in civic and religious life. Ballard was a druggist by trade, but in 1886 he became the police judge. He later relinquished that position to become postmaster. He served on the city council for over 30 years and is largely responsible for improving the city streets, sidewalks, and waterworks system. Ballard was a Mason, a director of the Lawrenceburg National Bank, and a Rotarian. He was also a supporter of a strong school system. In 1900, he was a delegate to the Republican National Convention. He and his wife, Gertrude Witherspoon, had two daughters. Ballard was known for his philanthropy, especially to disabled children. (Courtesy of Anderson County Historical Society.)

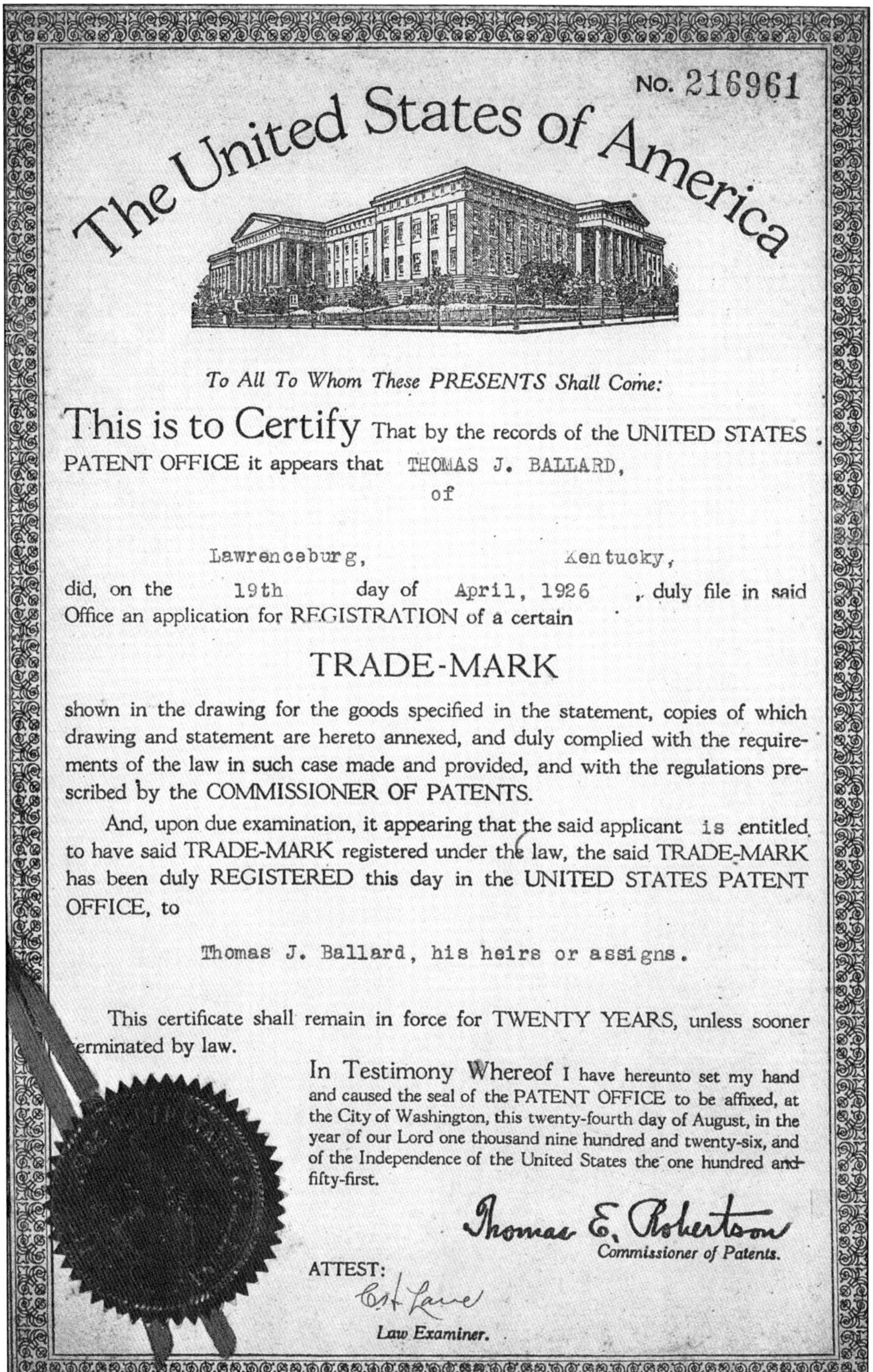

No. 216961

The United States of America

To All To Whom These PRESENTS Shall Come:

This is to Certify That by the records of the UNITED STATES PATENT OFFICE it appears that THOMAS J. BALLARD, of

Lawrenceburg, Kentucky,

did, on the 19th day of April, 1926, duly file in said Office an application for REGISTRATION of a certain

TRADE-MARK

shown in the drawing for the goods specified in the statement, copies of which drawing and statement are hereto annexed, and duly complied with the requirements of the law in such case made and provided, and with the regulations prescribed by the COMMISSIONER OF PATENTS.

And, upon due examination, it appearing that the said applicant is entitled to have said TRADE-MARK registered under the law, the said TRADE-MARK has been duly REGISTERED this day in the UNITED STATES PATENT OFFICE, to

Thomas J. Ballard, his heirs or assigns.

This certificate shall remain in force for TWENTY YEARS, unless sooner terminated by law.

In Testimony Whereof I have hereunto set my hand and caused the seal of the PATENT OFFICE to be affixed, at the City of Washington, this twenty-fourth day of August, in the year of our Lord one thousand nine hundred and twenty-six, and of the Independence of the United States the one hundred and fifty-first.

Thomas E. Robertson
Commissioner of Patents.

ATTEST:
E. H. Lane
Law Examiner.

This certificate is proof that Thomas J. Ballard received a trademark for his drug, Lawreline, a medicinal lotion for the treatment of sunburn and chapped lips. It also shows that Ballard continued to improve as a practicing druggist. (Courtesy of Geoghegan collection.)

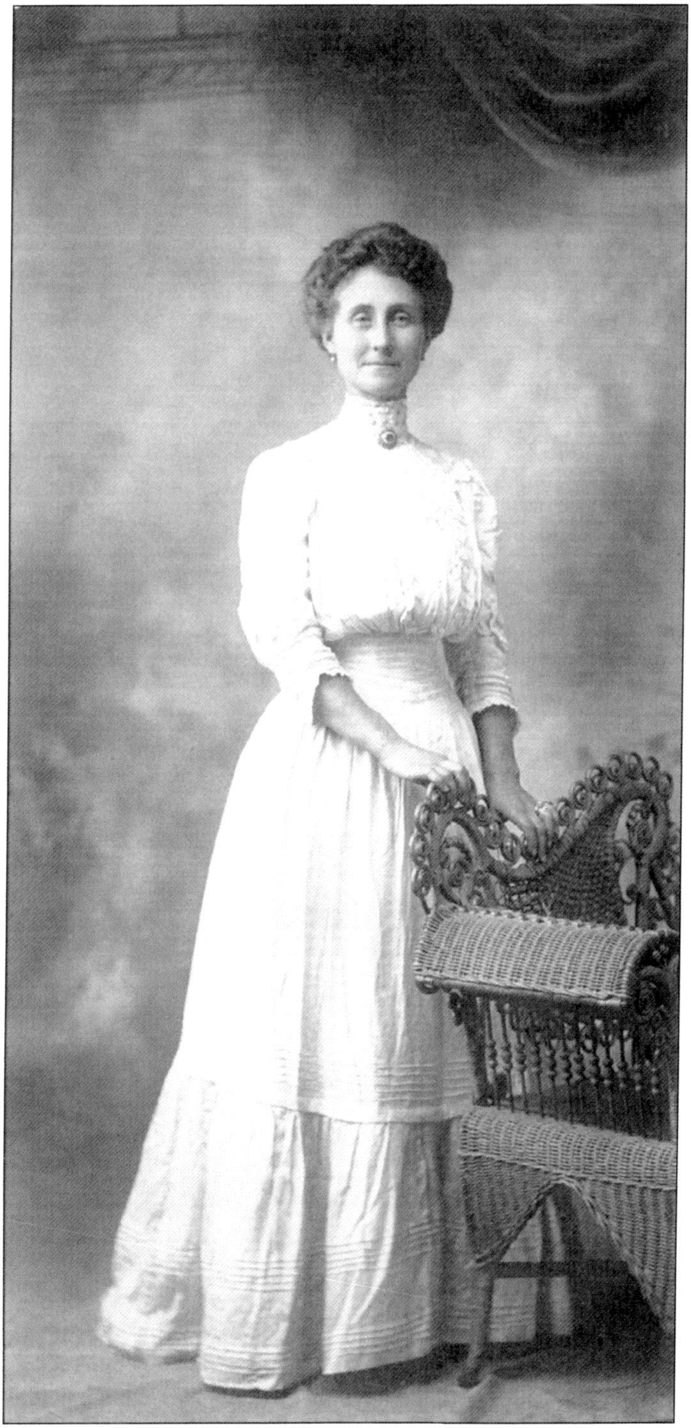

Gertrude Witherspoon Ballard was the second wife of Thomas J. Ballard. Her sister had been Ballard's first wife. Gertrude and Thomas had a wonderful life together, raising two daughters. Gertrude maintained an active social life and was involved in her church and community. (Courtesy of Geoghegan collection.)

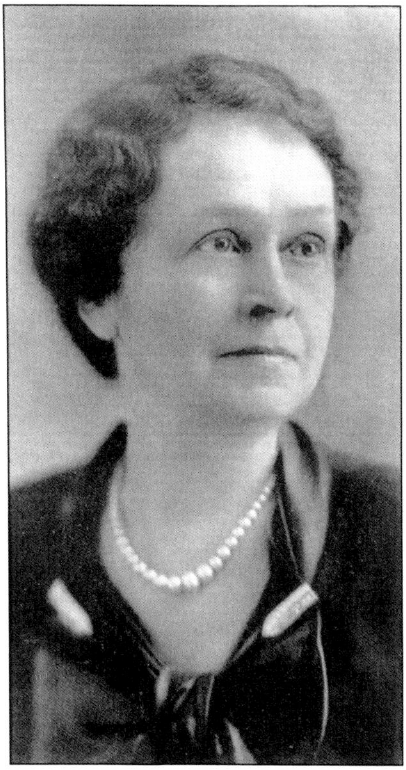

Lydia Kennedy Bond, a native of Mercer County, married W.T. Bond of the Bond family of Lawrenceburg. She was active in social and church affairs, wrote several pamphlets, and coauthored *The History of Anderson County* with L.W. McKee in 1937. (Courtesy of Anderson County Historical Society.)

Royal Oaks was the home of W.T. Bond. It was built in 1898 and burned in the 1970s. (Courtesy of Anderson County Historical Society.)

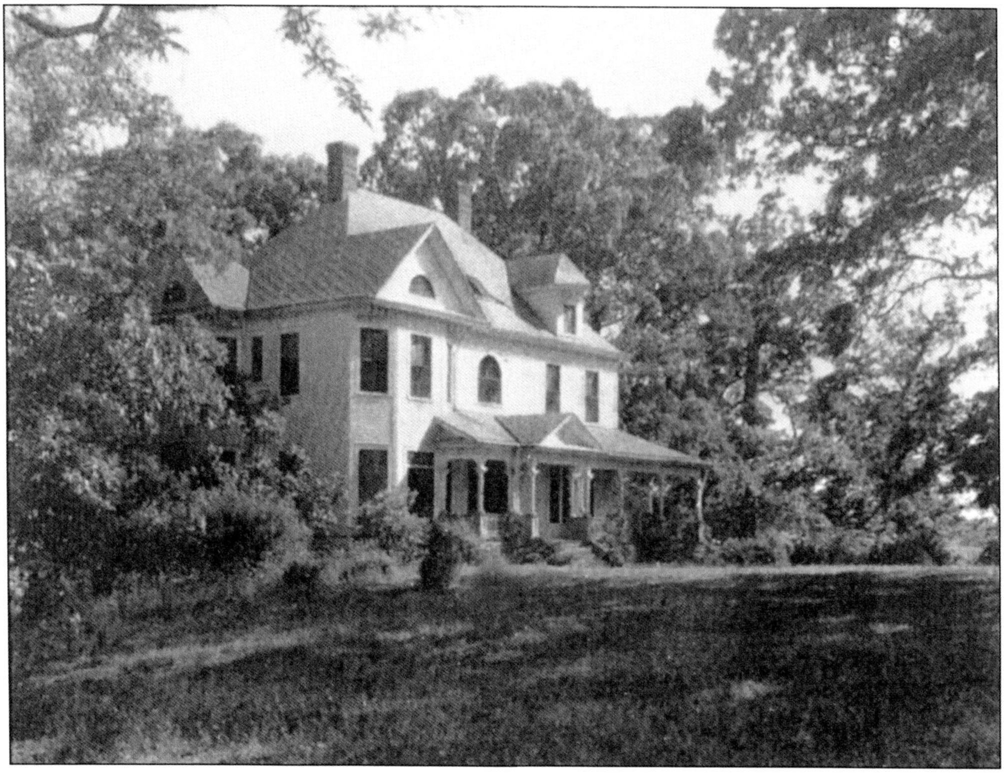

Dr. James Leslie Toll was one of the most civic-minded and respected men in Lawrenceburg. He attended Louisville Medical College while teaching school and graduated in 1897. He began practicing in Lawrenceburg in 1904. After the death of his first wife, he married Miss Fred Evans Filmore, and they had two daughters: Nancy Smith and Leslie Hollingsworth. He was a singer and was active in the First Christian Church, all while serving as mayor for 24 years. (Courtesy of McFarland collection.)

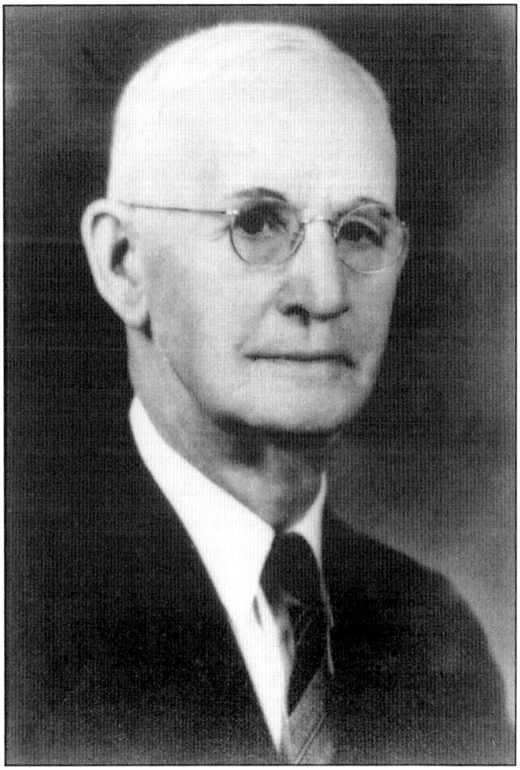

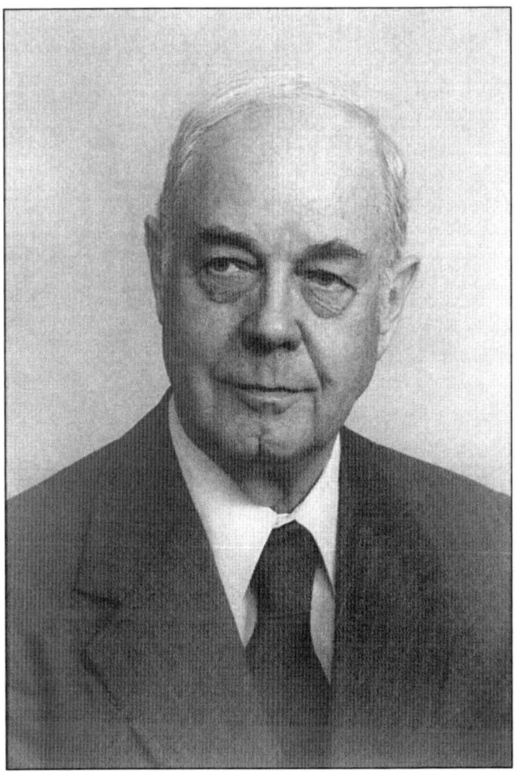

Ollie J. Bowen was orphaned as a child, but through good fortune he was able to lead an exemplary life. His early education was in Anderson County, and later he attended college and law school at George Washington University in Washington, DC. He returned to Lawrenceburg in 1933 and was elected to the state house of representatives and then the state senate in 1938 and 1940. He was the Anderson County attorney from 1940 until 1977, when he was elected district court judge. (Courtesy of Charlotte Bowen Wright.)

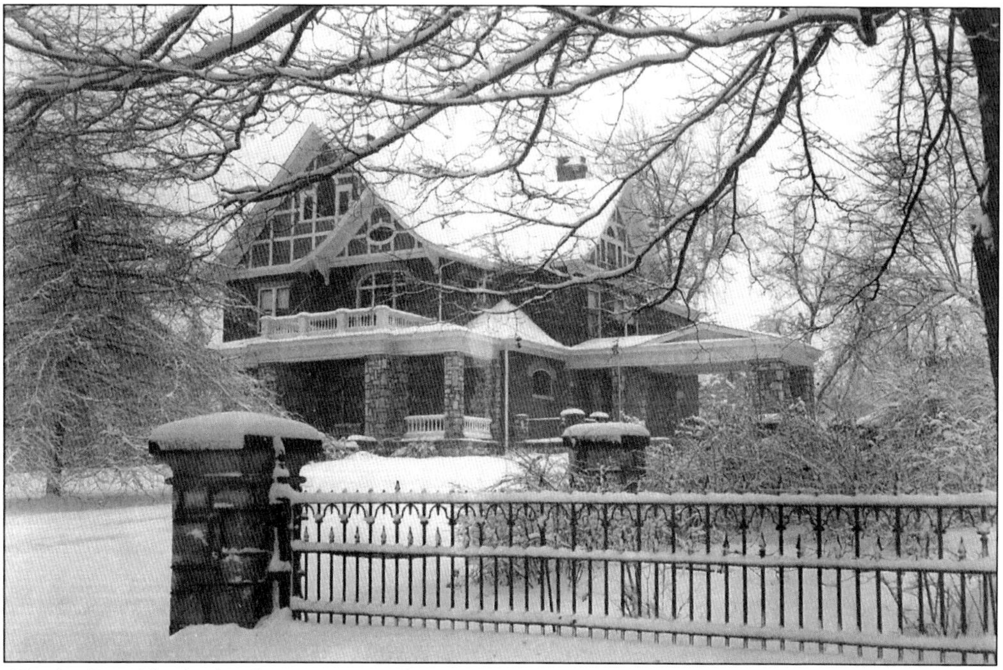

Charlie Bond built this home on South Main Street in 1860. Over the years, it has served as a residential home, a nursing home, and a funeral home. Today, it is the Gash Memorial Chapel. At the time of his death in 1917, Bond was president of the Lawrenceburg National Bank. He was also active in the real estate business; the building and development of Lawrenceburg were his delights. He built many homes and roads throughout central Kentucky. He was married to Ellen Witherspoon. (Courtesy of Anderson County Historical Society.)

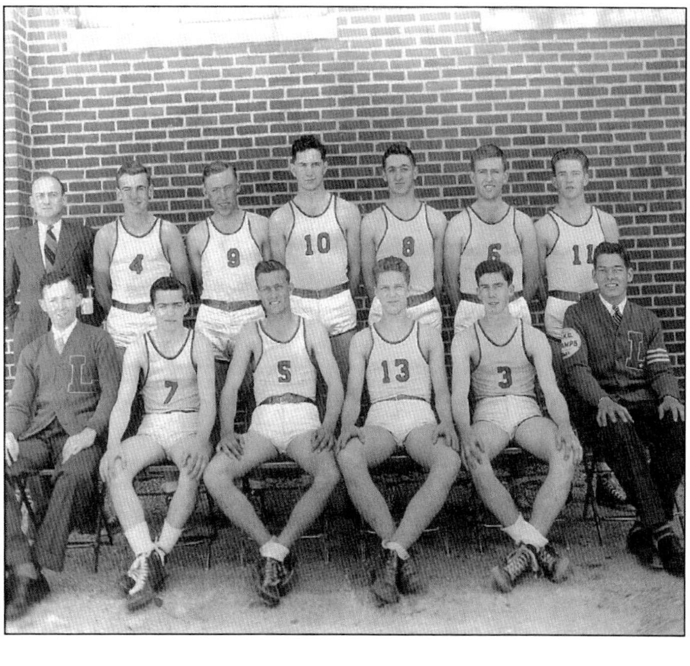

Edward Adams was a graduate of Lawrenceburg High School and the University of Kentucky. After college, he coached football and basketball and served as a teacher and principal at Lawrenceburg High School. His long career in education was followed by service in the Kentucky Fish and Wildlife Program. Adams (in his suit and tie in the back left) is seen here with his Lawrenceburg High School basketball team during the 1942–1943 season. (Courtesy of Birdwhistell collection.)

John William Gaines was born in Alton on May 22, 1869. He graduated from Kentucky Military Institute and taught in the Lawrenceburg Normal School for a brief time. In 1890, he was a clerk at the Anderson County Deposit Bank. Five years later, he was named vice president and manager. When the bank was reorganized as the Anderson National Bank, he became president. Gaines was admitted to the bar in 1909, served on the city council for eight years, and was mayor in 1921. (Courtesy of Anderson County Historical Society.)

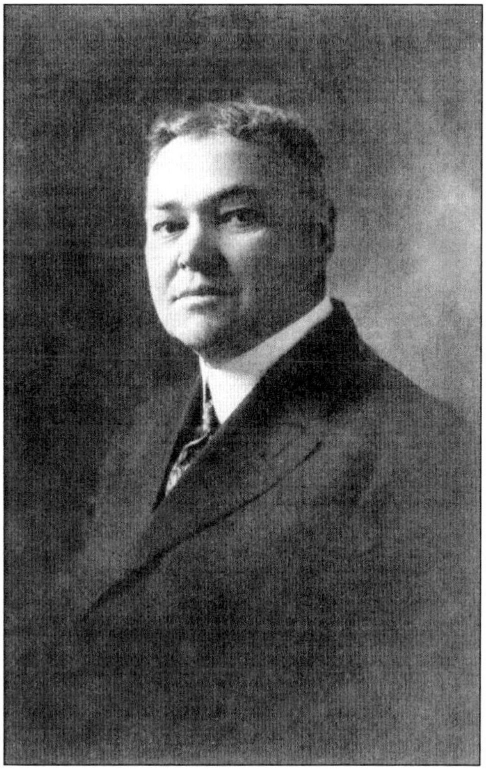

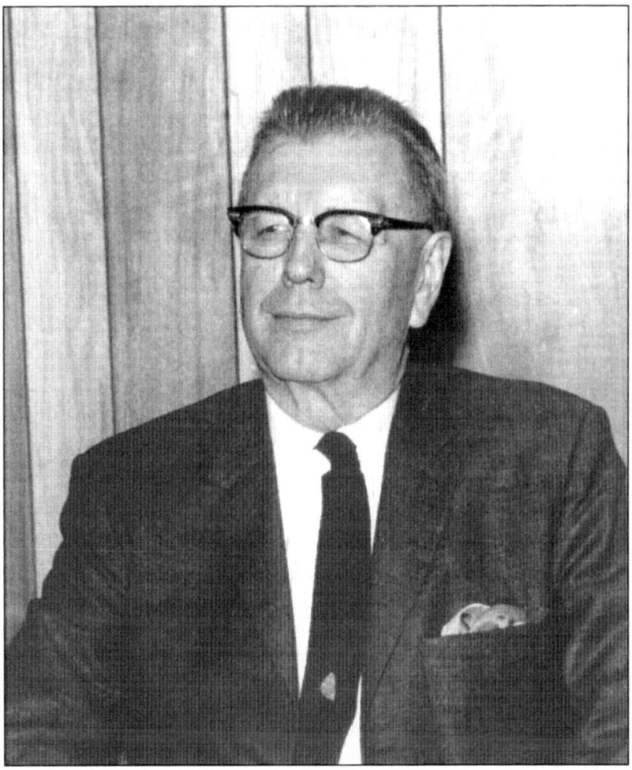

Frank Routt was vice president of the Anderson National Bank before serving as president from 1942 to 1970. Routt was a member of the Anderson County Industrial Board and served many civic endeavors. (Courtesy of Ripy-Crawford collection.)

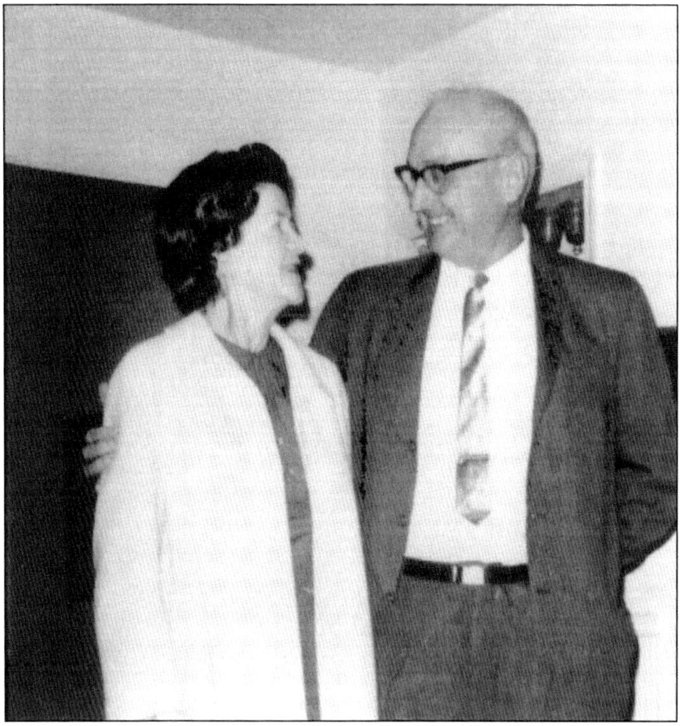

Walter Major Sr. was president of the Lawrenceburg National Bank after J.L. Sherwood. Major was a farmer and had been director of the bank prior to becoming president. His wife, Marjorie, was a sister to the McAfee brothers, who were famous horsemen in Anderson County. (Courtesy of Major collection.)

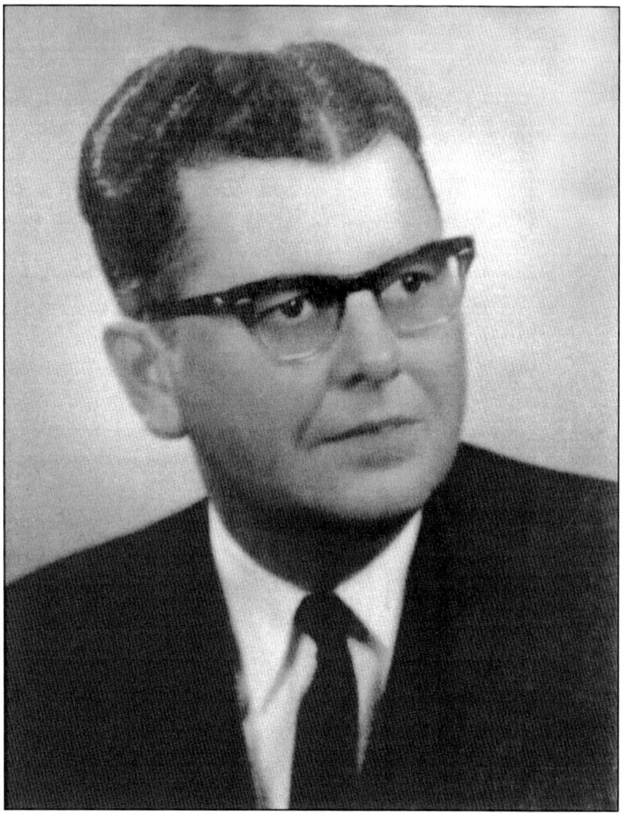

Marvin Edwards was a native of Mercer County but attended Lawrenceburg High School in order to play basketball. At Lawrenceburg High, he met his future wife, Edith Mae Hanks. After school, he and his father operated a small grocery store where they also made sausage—the beginnings of what became a large sausage enterprise. In 1962, Edwards was elected as a state senator. He eventually sold the sausage business to Webers and bought and ran the Travel World and Sports Center with his son. Marvin was a philanthropist known for always helping others. (Courtesy of Hawking collection.)

Walter Patrick was a native of Burgin, Kentucky. After serving his country in World War II, he returned home and entered the University of Kentucky, eventually graduating from University of Kentucky Law School and practicing in Lawrenceburg. Patrick served as the city attorney of Lawrenceburg and held other positions, including district judge. He was also involved in American Legion activities, especially those dealing with the Lawrenceburg Fair and Horse Show. (Courtesy of Barbara Patrick.)

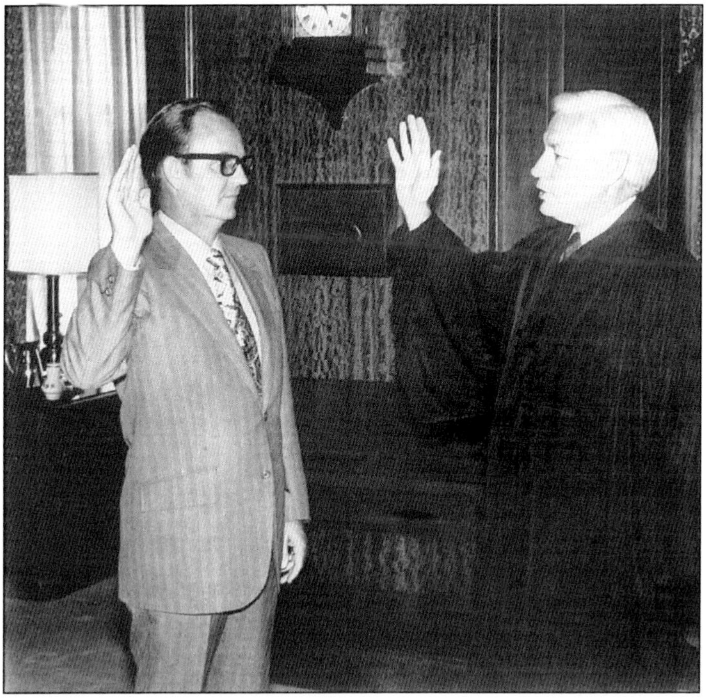

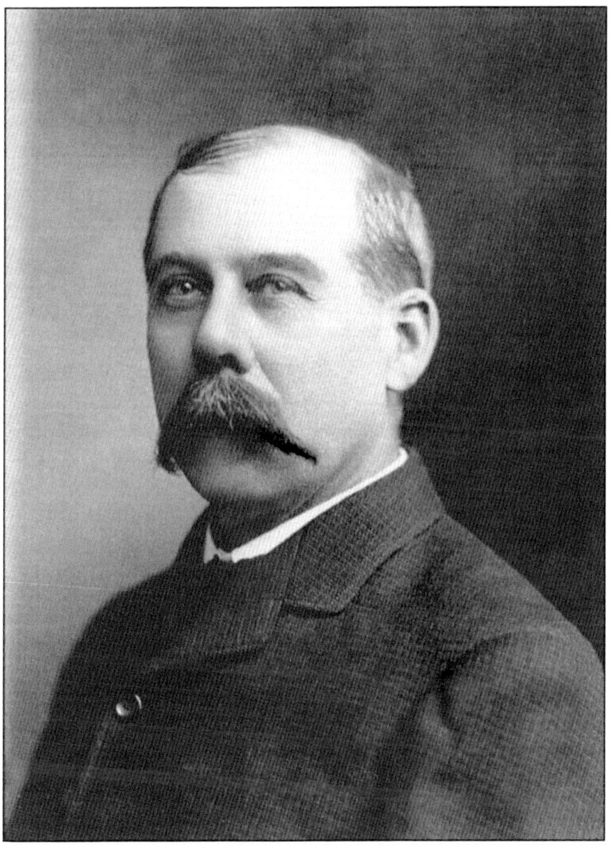

Jesse M. Johnson was a native of Indiana. After attending business college in Lexington, he and others organized the Lawrenceburg Bank in 1885. He was the cashier of the bank from its inception and later organized the Citizens Bank and Trust Company of Lawrenceburg, serving as its president until 1920, when it was consolidated with Lawrenceburg National Bank. He was the first elected mayor of Lawrenceburg and had interest in Standardbred trotting horses and the bourbon industry. (Courtesy of Ripy-Crawford collection.)

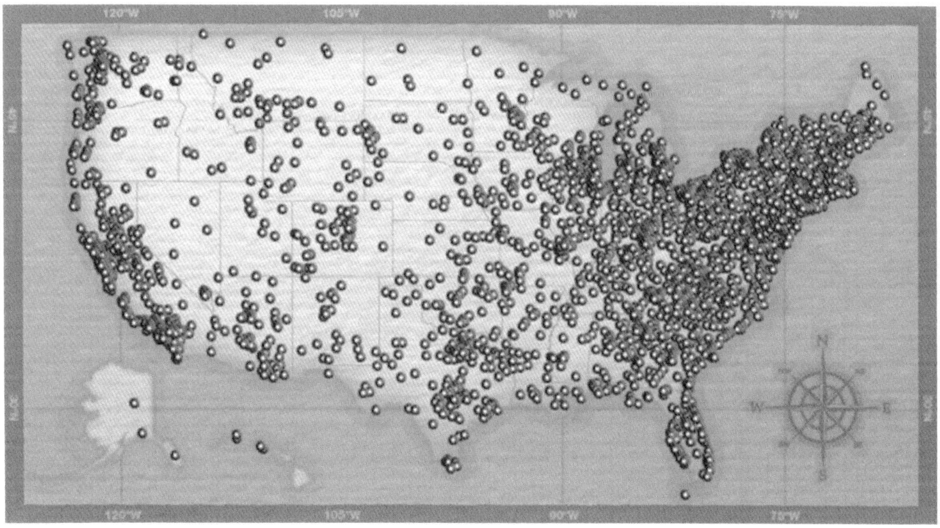